Preserving Rural Australia

KW-441-526

Editors: Alistar Robertson and Robyn Watts

Charles Sturt University

CSIRO
PUBLISHING

National Library of Australia Cataloguing-in-Publication entry

Preserving rural Australia: issues and solutions.

Bibliography.
ISBN 0 643 06388 9

1. Rural renewal – Australia. 2. Sustainable development – Australia. 3. Conservation of natural resources – Australia. 4. Natural resources– Australia – management. 5. Landscape protection – Australia. 6. Agricultural conservation – Australia. 7. Agriculture – Environmental aspects – Australia.
I. Robertson, A.I. (Alistar. I.). II. Watts, Robyn.

333.76160994

Published by
CSIRO PUBLISHING
PO Box 1139
Collingwood Vic 3066
Australia
Tel: (03) 9662 7666 Int: +61 3 9662 7666
Fax: (03) 9662 7555 Int: +61 3 9662 7555
Email: sales@publish.csiro.au
http://www..publish.csiro.au

Printed in Australia

Contents

Contributors   vii

Preface   ix

1   *Scientific and Social Impediments to Restoration Ecology*   1
    as Applied to Rural Landscapes
    Alistar I. Robertson and David A. Roshier

2   *The Farming Environment*   10
    Walter R. Stern and Ian C. McClintock

3   *Water, Politics and Power: Can We Integrate Natural Resource*   24
    Management in Rural Australia?
    Peter Cullen

4   *Challenges for Conservating Biodiversity in Australian*   33
    Freshwater Ecosystems
    Robyn J. Watts

5   *Water and Landscapes: Perceptions and Expectations*   43
    Kathleen H. Bowmer

6   *Sustaining Natural Resources and Biological Diversity in*   51
    Terrestrial Ecosystems of Rural Australia
    Tony Norton

7   *Nutrients and Algal Blooms: Lessons from Inland Catchments*   60
    Dhia Al Bakri and Mosharef Chowdhury

8   *Beefing up our Trade: Health and Environmental*   69
    Concerns and Rural Exports
    Marie Wynter

9   *Impediments to the Achievement of the Commercial and*   82
    Conservation Benefits of Farm Forestry
    Alan Black

10 *Toward Regional Strategies for Rural Sustainability:*   92
    A Farmer's View
    Peter Milliken

11 *Saline Politics: An Inland City Case Study*   100
    Petrina Quinn and Mark Conyers

12 *Growing Food and Growing Houses: Preserving Agricultural*   110
    Land on the Fringes of Cities
    Ian W. Sinclair

13 *Social and Economic Costs and Benefits of Taking Water*   125
    from our Rivers: the Macquarie Marshes as a Test Case
    Richard Kingsford

14 *Co-operative Management of Road Reserves for*   143
    Biodiversity Maintenance
    Quentin Farmar-Bowers

Contributors

Dr Dhia Al Bakri, Orange Agricultural College, The University of Sydney, PO Box 883, Orange, NSW 2800, Australia.

Professor Alan Black, Director, Centre for Social Research, Edith Cowan University, Joondalup, Western Australia 6027, Australia.

Professor Kathleen Bowmer, Deputy Vice-Chancellor (Academic), The Executive Centre, Charles Sturt University, PO Box 588, Wagga Wagga, NSW 2678, Australia.

Mr Mosharef Chowdhury, Orange Agricultural College, The University of Sydney, PO Box 883, Orange, NSW 2800, Australia

Dr Mark Conyers, Agricultural Research Institute, NSW Agriculture, PMB, Wagga Wagga, NSW 2650, Australia.

Professor Peter Cullen, FTSE, Director, Co-operative Research Centre for Freshwater Ecology, University of Canberra, PO Box 1, Belconnen, ACT 2616, Australia.

Dr Quentin Farmar–Bowers, ARRB Transport Research, 500 Burwood Hwy, Vermont South, Victoria 3133, Australia. Present address, Star Eight Consulting, 17 The Grange, East Malvern, Victoria, 3145, Australia.

Dr Richard Kingsford, NSW National Parks and Wildlife Service, PO Box 1967, Hurstville, NSW 2220, Australia.

Mr Ian C. McClintock, 'Milford Lodge', PO Box 118, Cootamundra, NSW 2590, Australia.

Mr Peter Milliken, 'Bertangles', Hay, NSW 2711, Australia.

Professor Tony Norton, The Johnstone Centre, Charles Sturt University, PO Box 789, Albury, NSW 2640, Australia. Present address: Department of Land Information, RMIT University, GPO Box 2476V, Melbourne, Victoria 3001, Australia.

Ms Petrina Quinn, School of Education, Charles Sturt University, PO Box 588, Wagga Wagga, NSW 2678, Australia.

Professor Alistar I. Robertson, The Johnstone Centre, School of Science and Technology, Charles Sturt University, PO Box 588, Wagga Wagga, NSW 2678, Australia.

Mr David A. Roshier, School of Science and Technology, Charles Sturt University, PO Box 588, Wagga Wagga, NSW 2678, Australia.

Mr Ian W. Sinclair, Manager Strategic Planning, Wollondilly Shire Council, PO Box 21, Picton, NSW 2571, Australia. Present address: EDGE Land Planning, PO Box 1858, Bowral, NSW 2576, Australia.

Emeritus Professor Walter R. Stern, Department of Plant Sciences, Faculty of Agriculture, The University of Western Australia, Nedlands, Western Australia 6907, Australia.

Dr Robyn J. Watts, The Johnstone Centre, School of Science and Technology, Charles Sturt University, PO Box 588, Wagga Wagga, NSW 2678, Australia.

Ms Marie Wynter, Centre for Resources and Environmental Studies, The Australian National University, Canberra, ACT 0200, Australia.

Preface

Scientific knowledge alone will not help Australia achieve sustainable management of land, water and biota. Scientists and the broader community now realise that a partnership is needed between land and water users, scientists, managers and the community if the nation is to achieve the goal of preserving rural resources. Such partnerships are not easy to achieve because of the differing perspectives bought to the issue of rural landscape management and the past lack of dialogue between rural communities and researchers.

The 'Rural Australia: Toward 2000' Conference held at Charles Sturt University in 1997 focused on the challenges facing rural Australians with regard to social, economic and ecological issues. The conference was organised in such a way that participants were challenged to think across disciplines when speaking about the impediments to sustainable rural communities. Participants included academics, rural administrators and rural residents from across Australia, New Zealand, Papua New Guinea and Thailand.

The conference was highly successful in its aim of getting members of the community and researchers from different disciplines to mix and converse. Several publications on various themes of the conference have already been produced by the Centre for Rural Social Research at Charles Sturt University.[1]

One of the themes of the conference was 'natural resources'. We accepted the challenge of acting as editors for a book which would be based on a selection of the papers delivered at the conference. Instead of purely technical papers, focused on narrow scientific problems, authors were asked to consider the social, cultural, political and economic dimensions of the natural resource management issue they dealt with.

The book contains a set of chapters dealing with broad issues relating to resource decline. A second set of chapters by farmers, rural town dwellers, a member of local government, and State and Federal resource managers, focuses on how different groups in the community deal with these issues to solve resource-use problems. The book illustrates the maturing of biophysical scientists in their understanding of the role of science in resource management issues and shows how the divide between investigator and knowledge users is being overcome in rural areas.

All chapters in the book have been subjected to peer review by between one and three referees. In the case of chapters by non-academics, such reviews mainly provided suggestions for making the contributions fit the style of this publication. In the cases of chapters written by professional scientists, the normal practice of critical appraisal of concepts and information was adhered to, and all chapters were modified following review.

Alistar Robertson and Robyn Watts

---

1   Doyle, R.U.(Ed.) (1998). *Rural Australia: Toward 2000 Conference keynote papers.* An edited collection of papers as presented at the conference (July 1997). (Centre for Rural Social Research: Wagga Wagga.); Alston, M. (Ed.) (1998). *Australian rural women towards 2000.* An edited collection of papers on women in rural Australia as presented at the Rural Australia: Toward 2000 Conference (July 1997). (Centre for Rural Social Research: Wagga Wagga.); Anscombe, A. W., and Doyle, R.U. (Eds) (1998). *Aboriginal and ethnic communities in rural Australia.* An edited collection of papers on Aboriginal and ethnic communities in rural Australia as presented at the Rural Australia: Toward 2000 Conference (July 1997). (Centre for Rural Social Research: Wagga Wagga.).

Scientific and social impediments to restoration ecology as applied to rural landscapes

Alistar I. Robertson and David A. Roshier

Introduction

There is much good news to tell about the restoration of rural landscapes in Australia. At the regional level, partnerships between government and resource users in the form of land and water management plans, and Landcare, Rivercare, and more recent projects under the umbrella of the National Heritage Trust, have contributed to changes in the management of biophysical resources on farms and in rural urban regions. Perhaps just as significantly, these shared projects have provided an avenue for social discourse on resource use and management (eg Curtis and Lockwood 1998; Milliken 1999; Quinn and Conyers 1999).

Scientific understanding of the ecological processes which sustain the potential productivity of the biosphere (as well as processes of degradation) has advanced dramatically in the last two decades. Agricultural scientists have demonstrated the benefits of new methods of farming, developed new crops and pasture management systems, and provided a series of tools to aid farmers with resource conservation. At the individual farm level, the use of whole farm planning, tree planting and fencing to reduce erosion and the adoption of more conservative approaches to tillage are all contributing to better use of resources and reduced soil losses (Pratley and Corbin 1994; Chan and Pratley 1998).

While it is true that changes in soil, water and vegetation management at the paddock and farm scale have decreased rates of soil erosion, maintained soil productivity and altered hydrological cycles for the better, this is too narrow a perspective from which to judge Australian agriculture as sustainable (Smith and McDonald 1998 and see below). Powerful indicators of system health at the landscape level, which focus on system function at landscape patch boundaries such as land–water margins (Rapport *et al.* 1998), suggest that farming systems in Australia are far from sustainable in a broader sense. Soil and water salinisation is increasing across catchments (Eberbach 1998), river system health is in decline (Harris and Gehrke 1997; Bowmer 1998) and there is continuing loss and fragmentation of bushland with subsequent loss of regional biodiversity (eg Department of the Environment, Sport and Territories 1995).

A hierarchy of linked economic, social and ecological factors are responsible for the continuing degradation of rural landscapes (State of the Environment Advisory Council 1996). Governments, research organisations and the community are now beginning to tackle these problems in a serious manner (eg Commonwealth of Australia 1992). One of the many positive outcomes of the focus on ecologically sustainable development has been the emergence of a new discipline, restoration ecology, as biophysical scientists have responded to society demands for better information on the methods to restore the ecological processes which underpin healthy landscapes (eg Jordon 1987; Lubchenco *et al.* 1991; Hobbs and Saunders 1993; Ludwig *et al.* 1997).

From the perspective of biophysical scientists who have researched land and water issues in regional Australia, it has become clear to us that notwithstanding (in fact, possibly owing to — see below) the significant structural changes in research and development policy and delivery in rural Australia (eg Curtis and Lockwood 1998), there remain major differences in the way that different groups of scientists, land and water managers and farmers perceive issues of sustainability in Australia. Many of the problems of communication among these groups has been discussed elsewhere (eg Copeland and Lewis 1997). Here we deal specifically with some of the scientific and social impediments to the application of restoration ecology to the rehabilitation of rural landscapes in Australia. We focus mainly on biophysical scientists and their interactions with farmers. The attitudes and behaviour of farmers and necessity for change have been discussed elsewhere (eg Vanclay 1995, Curtis and Lockwood 1998; Robertson and Pratley 1998).

Important issues for restoration ecology

There is no particular space or time scale that is the most relevant for managing ecosystems. The structure and 'behaviour' of any component of any ecosystem is a result of past and present factors operating across the landscape that surrounds them (Christensen *et al.* 1996). For instance, the salinity of river waters at any location in the Murray Darling Basin is determined by a hierarchy of factors. Past geological and climate events resulted in the development of the riverine plain over sedimentary layers containing salt. The interaction between variation in river flows, local geomorphology and groundwater inputs to rivers caused variation in river salinity prior to European settlement. Superimposed on this complex of factors, alterations to the spatial arrangement of vegetation and surface and groundwater flows at regional and local scales over the last 150 years have resulted in high salinity waters occurring more regularly and at more locations than previously.

Secondly, change is the normal course of events for most ecological systems, and at various scales ecosystems are composed of patches in different stages of recovery from natural disturbance (Wiens 1976). The science of ecology has moved away from concepts of equilibrium to embrace the view of dynamic ecosystems as it has become clear that natural disturbances have been and are the norm for ecosystems.

For rural Australian ecosystems there are some clear issues raised by the dynamic nature of ecosystems. Firstly, one of the most important resource management corollaries of disturbance to ecosystems is that ecosystems cannot be maintained indefinitely in the same state (Christensen *et al.* 1996). At very long time scales, climate and geologic changes have influenced, and continue to influence, the structure and function of global ecosystems. Humans have contributed to changes at continental and regional scales since prehistoric times. Therefore, concepts of 'naturalness' and 'the balance of nature' might have very limited time and space constraints (eg Taylor 1990). Setting appropriate goals for management and restoration projects is therefore dependent on an appreciation of time and space scales.

In addition, because humans are part of the landscape, the degree to which the landscape satisfies human needs and aspirations becomes an important component of evaluating ecosystem components (Lee 1992; Rapport *et al.* 1998) and setting restoration targets. Currently the development of restoration ecology as a practical science in Australian agricultural landscapes is limited by the inability of ecologists to understand or accept the alternative values for landscape elements held by primary producers and the general public.

Scientific impediments to restoration ecology

There are major gaps between the nature of ecological processes in Australian landscapes as understood by land and water managers and as understood by scientists (Morton 1996; Robertson and Pratley 1998; Roshier and Nichol 1998). This is illustrated by on-farm stocking rates that differ from those recommended by land management agencies, the lack of adoption by pastoralists of research results from studies directly applicable to their enterprise and environment (Morton 1996) and lack of agreement of farmers and scientists on water and remnant vegetation conservation (eg Barr and Cary 1992 ; Hobbs and Saunders 1993; Kingsford 1999). While some of these differences are related to socio-economic factors (see below), they also relate to the scales at which farmers and scientists study ecological and productivity information.

In the case of ecologists, much of our research may have been undertaken at spatial and temporal scales not relevant to the scale of resource usage by farmers. In the case of Australian rangelands, much of the early work by ecologists which showed a negative relationship between stocking rates and animal production did not adequately account for the scales of spatial and temporal variations in forage availability on pastoral properties (eg Ash and Stafford-Smith 1996; Roshier and Nicol 1998). The result has been severe scepticism among pastoralists towards assertions by scientists that reductions in stocking rate can result in higher productivity, because pastoralists manage their livestock at different scales to that examined by scientists.

Similar mismatches between the scales of scientific research and resource management exist for a variety of habitats and ecological processes. For instance, until comparatively recently most research on running waters in Australia was focused on small upland streams and rivers (eg see De Dekker and Williams 1986). However, many of the ecological problems arising from the damming of rivers to supply irrigation needs and drinking water occur downstream in lowland reaches of rivers and their associated wetland habitats. Signs of obvious stress in our lowland rivers began to emerge in the early 1990s, but even recent reviews of our knowledge of ecological processes in riverine systems (eg Lake 1995; Robertson *et al.* 1996) reveal that much of our understanding of these ecosystems relies on insights obtained from small streams, rather than from the large rivers themselves.

A major impediment to predicting the response of ecosystems to restoration is the lack of long-term ecological data sets. Our highly variable hydrological regime often means that the relatively short-term nature of our research projects (usually 3 years) causes us to miss the extreme event which controls the structure and function of ecosystems. Historical records held in sediments, tree rings and massive corals (eg De Dekker and Williams 1986; Isdale 1984) provide excellent records of long-term change, but do not reveal the mechanisms involved. There is no substitute for long-term data on ecological processes and their interactions over time in revealing the interplay of factors that shape ecosystems (Risser 1991).

Finally, biophysical scientists often perform their research in inappropriate habitats, with the result that research does not yield information appropriate for the restoration of landscapes. Many ecological studies are often focused on small patches of pristine habitats in the belief that understanding of ecological processes in undamaged sites will enable better restoration of altered sites. At the same time agricultural scientists often focus their research at the sub-paddock scale. The problem with both approaches is that the factors which control the organism or process under study may occur outside the study patch. Experience elsewhere indicates that scientists may only begin to ask the right questions about what factors are important in the structure and function of ecosystems when restoration is attempted (Jordon 1987). This indicates that ecological research on restoration or rehabilitation in rural Australia should focus on agricultural landscapes.

Socio-cultural impediments

Three of the main players in the management of the biophysical components of agricultural landscapes are farmers, management agency staff and scientists. Each of these groups has a distinct culture based on experience, training and environment (Beus and Dunlap 1992; Vanclay 1992, 1995). The perceptions of each group usually differ significantly with regard to the values they place on resources and their understanding of sustainability. Such differences represent major social impediments to setting targets for restoration projects.

Scientists often believe that primary producers aim to maximise profit and production, at the expense of resource maintenance. This is illustrated by significant differences in the responses of graziers and scientists to a question about the primary management objectives of beef producers in south-eastern Queensland (Figure 1.1). As well as maximising production and profit, beef

Figure 1.1   Graziers' and research scientists' perceptions of the primary management objective of the majority of beef producers in Queensland. Based on responses from 81 producers and 30 research scientists. Data from MacLeod and Taylor (1994).

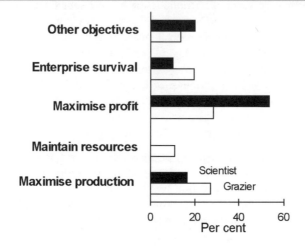

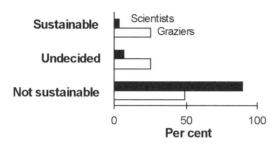

Figure 1.2   Graziers' and research scientists' perceptions of the sustainability of present grazing practices. Summary of data in MacLeod and Taylor (1994).

producers identified enterprise survival and resource maintenance as being important objectives for most members of their industry. While there are no similar data for other groups of scientists, our experience with producers (eg see Milliken 1999) and a variety of other surveys indicate a strong stewardship ethic among many Australian primary producers (Vanclay 1997).

Many scientists consider that current agricultural practices are not sustainable, while the perceptions of producers are often more variable. Again, scientists and graziers in Queensland illustrate this point. When asked whether they considered the beef grazing practices that were commonly adopted in Queensland to be sustainable, there were significant differences between the two groups (Figure 1.2). Interestingly, nearly 50% of producers considered current practices to be not sustainable.

Both producers and scientists agreed that excessive stocking rates, limited information and knowledge and adverse weather conditions were the main factors responsible for causing land degradation problems, and the majority of both groups thought that it was technically feasible to rectify land degradation (MacLeod and Taylor 1994). The differences in cultures became most apparent when both groups were asked which groups held, or were in a position to generate, knowledge appropriate to developing sustainable systems of grazing land management (Figure 1.3). Both groups thought that the necessary knowledge was held by a limited range of people, and both groups strongly favoured themselves as principle sources of knowledge. Also, graziers often thought that groups other than scientists were more likely to possess the necessary knowledge.

Figure 1.3   Graziers' and research scientists' perceptions of the group which might possess knowledge appropriate to developing sustainable systems of grazing land management. Data from MacLeod and Taylor (1994).

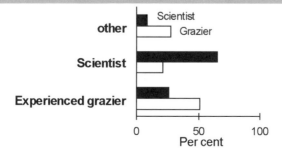

Interestingly, Landcare groups, professional management consultants, or agribusiness groups were not mentioned within the category 'other' (Figure 1.3), while professional extension officers were nominated by only 9% of beef producers as likely to have the appropriate knowledge.

These data (Figure 1.3) reflect the responses of primary producers in other surveys about land management (eg Bennett *et al.* 1997 and Prof. J. Bennett, personal communication), where landowners feel that because they have the day-to-day experience managing resources, they are the most likely to possess the knowledge required for restoration. Given our experience with the scepticism that both primary producers and scientists have about each other's expertise in sustainable resource management, we take heart from the survey (Figure 1.3), because it reveals that a significant percentage (more than 20%) of both groups acknowledge that they are both key players in potential strategies for restoring agricultural landscapes.

Among the more progressive scientists and farmers there has been a shift to new ways of mixing production and conservation imperatives in agricultural landscapes (eg Klomp and Lunt 1997; Hale and Lamb 1997). However, most farmers and scientists have a long way to go before they change their attitudes to conventional agriculture. For scientists, major cultural differences among scientists involved in research on agricultural landscapes is impeding a more rapid shift to alternative agricultural practices, and the conservation of native biota and ecological processes in rural landscapes.

Recent research on the attitudes of faculty members in Land Grant Universities in the United States indicates that academics from different disciplines have very different attitudes to conventional and alternative agriculture (Beus and Dunlap 1992; Lyson 1998). Academics are often more conventional than farmers, and far more conventional than known proponents of alternative agriculture, when asked about their attitudes to conventional and alternative agriculture (Beus and Dunlap 1992). In addition, academics from more natural resource science disciplines tend to most strongly

Figure 1.4   Responses of faculty members from different discipline groups in the School of Agriculture and Home Economics at Washington State University to a survey of their attitudes to alternative and conventional agriculture. Higher scores on the ACAP (alternative–conventional agriculture paradigm) scale denote stronger endorsement of alternative agriculture (eg minimum tillage, reduced chemical use, intercropping). Based on Beus and Dunlap (1992).

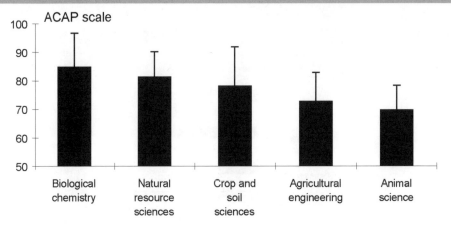

endorse alternative agriculture, while academics from more production oriented departments are the most conventional (Figure 1.4). No similar local data exist, but it would be no surprise if a survey of Australian academics produced similar results.

Factors such as age, sex and background (urban or agricultural) are important contributors to such differences (Beus and Dunlap 1992; Lyson 1998), but in our experience 'capture' of academics by agribusiness or conservation group cultures plays a major role in the attitudes and behaviour of scientists. With regard to conservation of native biota in Australian agricultural landscapes, we believe that a large number of ecologists lack an appreciation of the economic and social realities of farming. This manifests itself in various ways, ranging from failure to keep landowners informed of the outcome of ecological work carried out on private properties through to insistence that the only way to ensure biodiversity conservation is to promote reserves at the exclusion of agricultural production. In the latter case, ecologists may be 'captured' by the conservation lobby.

For agricultural scientists the most obvious examples of 'capture' are those whose research is funded directly by agribusiness. However, Federal Government science policy that has promoted co-operative research ventures focused on particular agricultural industries, such as the Cooperative Research Centres for Sustainable Rice Production and Sustainable Cotton Production, has helped to capture scientists within the particular focus of these industries. While such arrangements will yield more sustainable farming practices at the paddock and farm scale, they are unlikely to be arenas for consideration of landscape scale issues of land and water utilisation or the conservation biology of native biota, owing to their narrow focus.

The next 50 years

Following a period of readjustment between 1970 and the present, Australian agriculture is likely to enter another phase of productivity like that seen in the period after the Second World War. The world population is growing rapidly and, notwithstanding the present economic malaise in many parts of the world, demand for Australian produce will rise. Secondly, the greater efficiency of farmers, based on better farm management, precision farming, more direct marketing of produce, and economy of scales as we shift to more and larger corporate farms, will enable Australian primary producers to better penetrate global markets. In addition, the biotechnology revolution is now providing and will continue to provide new products with higher yields that are better suited to specific environmental conditions. Finally, there is a growing trend for greater corporate investment in agribusiness ventures by fund managers.

While such an expansion will be good for the economy of the country in the short-term, it will provide a major challenge to the sustainability of our already stressed agroecosystems because there will be greater demand on water resources, more intensive use of soil and greater pressure on vegetation. Scientists are still grappling with the changes in research scope and content required for current ecosystems management. Later in this volume Cullen (1999) lists four major challenges facing the sustainable management of natural resources: greed, ignorance, institutional arrangements and fashions in government. To this we would add the attitudes and behaviour of biophysical scientists and farmers.

A better understanding of the motivation of resource scientists will be a necessary first step to removing some of the impediments to the application of restoration ecology in rural landscapes. It will also be a precursor to designing tertiary level courses for the next generation of scientists who will have to deal with the challenges posed by an expanding agricultural sector. An important part of future tertiary courses for scientists will be the incorporation of training in the social sciences, so that science graduates have the skills to 'place' their work in a context likely to be meaningful to primary producers.

Finally, biophysical scientists may remove some of the scientific impediments to the application of restoration ecology as a useful discipline by establishing long-term ecological research sites and programs. Such sites will need to include the mosaic of habitats which make up rural Australia — farms, rivers and associated habitats and woodland or forest remnants. Partnerships between primary producers, scientists and management agency staff in ecosytem management on the different space and time scales relevant to farmers and ecological processes is the only way to restore rural landscapes.

References

Ash, A.J., and Stafford-Smith, D.M. (1996). Evaluating stocking rate impacts in rangelands: animals don't practice what we preach. *Rangelands Journal* **18**, 216–243.

Barr, N., and Cary, J. (1992). *Greening a brown land.* (MacMillan Education Australia Pty Ltd: Melbourne.)

Beus, C.E., and Dunlap, R.E. (1992). The alternative–conventional agriculture debate: where do agricultural faculty stand. *Rural Sociology* **57**, 363–380.

Bennett, J., Blamey, R., and Morrison, M. (1997). Valuing damage to South Australian wetlands using the Contingent Valuation Method. Occasional Paper No 13/97, LWRRDC.

Bowmer, K. H. (1998). Water: quantity and quality. In *Agriculture and the environmental imperative.* (Eds J.E. Pratley and A.I. Robertson.) (CSIRO Publishing: Melbourne.)

Chan, K.Y., and Pratley, J.E. (1998) Soil structural decline: How can the trend be reversed? In *Agriculture and the environmental imperative.* (Eds J.E. Pratley and A.I. Robertson.) (CSIRO Publishing, Melbourne.).

Christensen, N.L., Bartuska, A.M., Brown, J.H., Carpenter, S., D'Antonio, C., Francis, R., Franklin, J.R., McMahon, J.A., Noss, R.F., Partson, D.J., Peterson, C.H., Turner, M.C., and Woodmansee, R.G. (1996). The report of the Ecological Society of America committee on the scientific basis for ecosystem management. *Ecological Applications* **6**, 665–691.

Commonwealth of Australia (1992). *National strategy for ecologically sustainable development.* (Australian Government Publishing Service: Canberra.)

Copeland, C., and Lewis, D. (Eds) (1997). *Saving our natural heritage? The role of science in managing Australia's ecosystems.* (Halstead Press: Sydney.)

Cullen, P. (1999). Water, politics and power: can we integrate natural resource management in rural Australia? In *Preserving rural Australia: issues and solutions.* (Eds A.I. Robertson and R. Watts.) (CSIRO Publishing, Melbourne.)

Curtis, A., and Lockwood, M. (1998). Natural resource policy for rural Australia. In *Agriculture and the Environmental Imperative.* (Eds J.E. Pratley and A.I. Robertson.) (CSIRO Publishing: Melbourne.)

De Dekker, P., and Williams W.D. (Eds) (1986). *Limnology in Australia.* (CSIRO and Dr W Junk: Canberra.)

Department of Environment, Sport and Territories (1995). *Native vegetation clearance, habitat loss and biodiversity decline.* Biodiversity Series, Paper No. 6. (Biodiversity Unit, Department of Environment, Sport and Territories: Canberra.)

Eberbach, P.L. (1998). Salt-affected soils: their causes, management and cost. In *Agriculture and the environmental imperative.* (Eds J.E. Pratley, and A.I. Robertson.) (CSIRO Publishing: Melbourne.)

Hale, P., and Lamb, D. (Eds) (1997). *Conservation outside nature reserves.* (Centre for Conservation Biology, University of Queensland: Brisbane.)

Harris, J.H., and Gehrke, P.C. (Eds) (1997). *Fish and rivers in stress.* The NSW rivers survey. (NSW Fisheries Office of Conservation and the Cooperative Research Centre for Freshwater Ecology: Cronulla.)

Hobbs, R.J., and Saunders, D. (Eds) (1993). *Reintegrating fragmented landscapes. Towards sustainable production and nature conservation.* (Springer Verlag: New York.)

Isdale, P. (1984). Fluorescent bands in massive corals record centuries of coastal rainfall. *Nature* **310**, 578–579.

Jordon, W.R. (Ed.) (1987). *Restoration ecology: a synthetic approach to ecological research.* (Cambridge: New York.)

Kingsford, R. (1999). Social and economic costs and benefits of taking water from dryland rivers: the Macquarie Marshes as a test case. In *Preserving rural Australia: issues and solutions.* (Eds A.I. Robertson and R.J. Watts) (CSIRO Publishing, Melbourne.)

Klomp, N., and Lunt, I. (Eds) (1997). *Frontiers in ecology: building the links.* (Elsevier: Oxford.)

Lake, P.S. (1995). Of floods and droughts: river and stream ecosystems of Australia. In *Rivers and stream ecosystems, Ecosystems of the world 22*. (Eds C.E. Cushing, K.W. Cummins and G.W. Minshall.) pp. 659–694. (Elsevier: Amsterdam.)

Lee, R.G. (1992). Ecologically effective social organisation as a requirement for sustaining watershed ecosystems. In *Watershed management*. (Ed. R.J. Naiman.) pp. 73–90. (Springer: New York.)

Lubchenco, J., Olsen, A.M., Brubaker L.B., Carpenter S.R., Holland, M.M., Hubbell, S.P., Levin, S.A., McMahon, J.A., Matson, P.A., Mellilo, J.M., Mooney, H.A., Peterson, C.H., Pulliam, H.R. Real, L.A., Regal P.J., and Risser, P.G. (1991). The sustainable biosphere initiative: an ecological research agenda. *Ecology* **72**, 371–412.

Ludwig, J., Tongway, D., Freudenberger, D., Noble J., and Hodgkinson, K. (Eds). (1997). *Landscape ecology, function and management. Principles from Australia's rangelands*. (CSIRO Publishing: Melbourne.)

Lyson, T.A. (1998). Environmental, economic and social aspects of sustainable agriculture in American land grant universities. *Journal of Sustainable Agriculture* **12**, 119–130.

MacLeod, N.D., and Taylor, J.A. (1994). Perceptions of beef cattle producers and scientists relating to sustainable land use issues and their implications for technology transfer. *Rangelands Journal* **16**, 238–253.

Milliken, P. (1999). Toward regional strategies for rural sustainability — a farmers involvement. In *Preserving rural Australia: issues and solutions*. (Eds A.I. Robertson and R. Watts.) (CSIRO Publishing: Melbourne.)

Morton, S.R. (1996). Land management and population ecology. In *Frontiers of population ecology*. (Eds R.B. Floyd, A.W. Shepard and P.J. de Barro.) pp. 509–529. (CSIRO Publishing: Melbourne.)

Quinn, P., and Conyers, M. (1999). Saline politics: Problem description and local involvement. In *Preserving rural Australia: issues and solutions*. (Eds A.I. Robertson, and R.J. Watts.) (CSIRO Publishing: Melbourne.)

Pratley, J. E., and Corbin, E.J. (1994). Cultural practices. In *Principles of field crop production*. (Ed J.E. Pratley.) pp. 302–348. (Oxford University Press: Sydney.)

Rapport, D.J., Gaudet, C., Karr, J.R., Baron, J.S., Bohlen, C., Jackson, W., Jones, B., Naiman, R.J., Norton, B., and Pollock, M.M. (1998). Evaluating landscape health: integrating societal goals and biophysical processes. *Journal of Environmental Management* **53**, 1–15.

Risser, P.G. (Ed.) (1991). *Long-term ecological research. An international perspective*. (John Wiley and Sons, Chichester.)

Robertson, A.I., and Pratley, J.E. (1998). From farm management to ecosystem management. In *Agriculture and the environmental imperative*. (Eds J.E. Pratley and A.I. Robertson.) (CSIRO Publishing: Melbourne.)

Robertson, A.I., Boon, P., Bunn, S., Ganf, G., Herzceg, A., Hillman, T., Walker, K.F. (1996). A scoping study into the role, importance, sources, transformations and cycling of carbon in the riverine environment. Final Report on Natural Resource Management Strategy Project No. R6067. (Murray Darling Basin Commission: Canberra.)

Roshier, D.A., and Nicol, H.I. (1998). Implications of spatio-temporal variation in forage production and utilisation for animal productivity in extensive grazing systems. *Rangelands Journal* **20**, 3–25.

Smith, C.S., and McDonald, G.T. (1998). Assessing the sustainability of agriculture at the planning stage. *Journal of Environmental Management* **52**, 15-37.

State of the Environment Advisory Council (1996). *Australia: state of the environment*. (Environment Australia: Canberra.)

Taylor, S.G. (1990). Naturalness: the concept and its application to Australian ecosystems. In J.E. Pratley and A.I. Robertson (Eds). *Australian ecosystems: 200 years of utilisation, degradation and reconstruction*, Proceedings of the Ecological Society of Australia 16, pp. 411–418.

Vanclay, F. (1992). Farmer attitudes or media depiction: which is the barrier to adoption? *Regular Journal of Social Issues* **26**, 41–50.

Vanclay, F. (1995). Actors and structures in agriculture. In *With a rural focus*. (Ed. F.Vanclay.) pp.411–418. (Centre for Rural Social Research, Charles Sturt University: Wagga Wagga.)

Vanclay, F. (1997). The sociological context of environmental management in agriculture. In *Critical landcare*. (Eds S. Lockie, and F.Vanclay.) pp. 9–28. (Centre for Rural Social Studies, Charles Sturt University: Wagga Wagga.) Key Papers Series Number 5.

Wiens, J.A. (1976). Population responses to patchy environments. *Annual Review of Ecology and Systematics* **7**: 81–120.

2 The farming environment

Walter R. Stern and Ian C. McClintock

'… if we are to reap the harvest of this biotechnological revolution it will have to be predicated on another: a major change in the way we think about, organise, capitalise and carry out farming, processing and marketing …' (H.W.Woolhouse 1994)

Introduction

Within our region and beyond it, Australia is an exporter of agricultural products and there is also a significant trade in food within Australia.

Over the last 25 years, changes in the speed of communication, travel and transport, and the formation of trading blocks have altered the pattern of agricultural trade. At the same time there have been major changes in agricultural practices. Because of recent advances in understanding biological and ecological processes, the way in which agricultural research is conducted and organised has also changed markedly. We are not immune to climatic hazards, pests, diseases and environmental degradation and these have a significant influence on the production process. The interaction of the many factors that go to make up the agricultural industries affects the farming environment, now and in the foreseeable future.

Here we address many topics relating to farm management. These are principally technical and technological matters but inevitably we touch on aspects of the environment, agricultural economics and sociology. We do not claim special expertise in these areas, and in this chapter we have taken a broad sweep rather than a narrow focus. We pose questions rather than offer prescriptions and this is deliberate. We wish to highlight the many factors — some contradictory and some unknown — that farmers or groups of farmers may face, when they have to make short- or long-term management decisions. Our approach is eclectic. We hope this contribution will stimulate discussion of some issues that are important to the farming community.

In this chapter we first establish a framework and follow that up with a definition of farming systems that embraces concepts of sustainable agriculture. We then explore some issues that might impact on farming systems and individual primary producers. We also touch on the role and value of research in the future, and on the importance of maintaining services to the farming community.

Framework

The terms of trade, defined as 'the ratio of prices received for outputs to the prices paid for inputs', have been declining steadily for Australian farmers for many years, yet agricultural output has continued to grow because of steady improvements in productivity (Wonder and Fisher 1990). This growth in productivity has been made possible through the release of improved crop cultivars, livestock breeding and selection, judicious use of chemicals and fertilisers, more energy efficient machinery and equipment, and innovations in farm practices. Yields of the major crops have been rising steadily (Russell 1973; Perry and D'Antuono 1989) but this has not been uniformly so (Hamblin and Kyneur 1993). There is great variability in yield between and within agroecological areas. In non-irrigated areas, average agricultural production is closely associated with rainfall over most of the area, and this has been demonstrated for wheat by French and Schultz (1984). However, where rainfall is low, as at the drier margins or in areas where it is markedly irregular and variable (Fitzpatrick 1964, 1970), yields may not be sustained (Hamblin and Kyneur 1993). An important factor contributing to yield variation between farms is the management skill of individual farmers, who juggle the resources at their disposal within the prevailing environmental and marketing constraints.

Within an area, the dominant activity by farmers determines the type of farming system for that area.

Farming systems

The following statement provides a definition of farming systems, as well as expressing significant elements of the concept of sustainable agriculture:

> 'Farming systems are based on the productive capacity of the land, the availability of suitable varieties or breeds, the constraints imposed by such factors as pests and diseases and the probability of occurence of natural hazards. They are usually shaped by the physical environment and the availability of resources, by the prevailing economic conditions, and by the inclination and capacities of primary producers. Farming systems also arise out of a capacity by the community to service farms and farmers, and to keep up the supply of ingredients for production such as suitable new varieties and breeds, fertilizers and chemicals, appropriate machinery and fuel, finance, communication and transport, and the capacity to market efficiently or value-add to the farm product. Farming systems have evolved so as to secure profits for individual farms, while maintaining the productivity of the land. If any part of the system should fail, the farming system may break down or may change. Proof that a farming system is successful is found in the economic viability of the farmers that adopt the system while sustaining the productivity of the land.' (modified from Stern 1979)

Farming systems will vary with agroecological regions and with the type, scale and intensity of enterprise. The agroecological areas of Australia are depicted in Figure 2.1 of *Australian agriculture and the environment* (McLennan 1996, figure 1.5). The evolution of farming systems in Australia can be traced in the economic analyses undertaken by Davidson (1981) and these provide some interesting perspectives.

Figure 2.1   The principal agroecological regions of Australia, as portrayed by McLennan 1996. (The original source was from the Standing Committee on Agriculture, 1991).

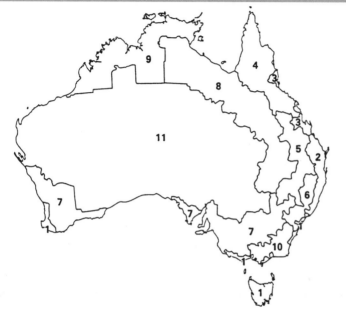

1   Wet Temperate Coasts

2   Wet Sub-Tropical Coast

3   Wet Tropical Coast and Tableland

4   North-East Wet/Dry Tropics

5   Sub-Tropical Slopes and Plains

6   Sub Tropical Highlands

7   Temperate Semi-Arid Slopes and Plains

8   Semi-Arid Tropical and Subtropical Plainlands

9   North-Western Wet/Dry Tropics

10 Temperate Highlands

11 Arid Interior

There is no agreed measure of 'technological progress' in farming systems over time, but it could be argued that such measures as fertiliser use, chemical use, power and energy consumption in operating equipment and machinery, water use efficiencies and output/unit of labour, or a combination of these, may be indicators of change. Water quality is a good monitor of the state of health of the resource base (Walker and Reuter 1996); physical and chemical measures of soil attributes could also be used as indicators of change in land management (Larson and Pierce 1994).

Resources

Types of resources

Resources are managed by people — either as individuals or in groups or organisations. These may be environmental or aggregate, and are differentiated below.

Environmental resources such as land, soil and water (including rainfall) and *genetic resources* of plants and animals are those which have been traditionally associated with agriculture. Individual landholders have control over some of these and not others. As we enter the next millennium, further clearing of land for agriculture will no longer be possible. Pressure to use the available land more intensively will increase, and, unless farming systems have been developed to cope with this, undesirable environmental consequences may ensue. Community pressures for environmental protection may impose additional demands, and sometimes serious constraints, on farming. For example, in NSW, legislation is under consideration concerning threatened species, regional vegetation plans and the cessation of clearing. The arid or rangelands areas of the continent present special problems in environmental management that transcend resource use and conservation, and extend to sociological issues.

Aggregate resources are those shared by society and include public bodies as well as private organisations. They provide the infrastructure for the agricultural industries and enable individual primary producers to function at a distance, nationally and globally. These resources include financial institutions — *monetary resources*; educational establishments, research and biotechnology organisations or firms, bodies offering information and communication (this includes the Internet) — the *information resource*; the *transport resource* (land, sea and air) and, last but not least, the existence of marketing organisations — *intelligence resource*.

Primary producers and their organisations either lobby for, contribute to, or participate in developing and maintaining many of these resources. Such infrastructures are vital for the functioning and well-being of agriculture and their existence and accessibility contribute significantly to the efficiency of the agricultural sector as a whole.

The way individual primary producers tap into and are able to use these resources contributes to the success of their individual enterprise and of the well-being of the resource. In aggregate terms, the success of farming systems rests on how effectively groups of primary producers secure and deploy the resources they require.

Farm practices and the maintenance of resources

The period of expansion and exploitation in the 1890s was based on traditional European farm practices of the time, and on the grazing of native grasses and forage plants. During the first half of the twentieth century, large areas of Australia have benefited from sound fertiliser practices. Soil fertility has been built up with regular fertiliser dressings, principally superphosphate, on soils of poor nutrient status (Stephens and Donald 1958); application of trace elements and forage legumes has brought previously unproductive land into production, especially in South Australia and Western Australia (Donald and Prescott 1975). After the first and second world wars, subdivision of large holdings and the clearing of land for soldier settlement led to closer settlement and the wider adoption of introduced and more productive pasture species and improved crop types. This period also coincided with rapid farm mechanisation, especially after the invention of the three-point linkage on tractors and the subsequent development of new types of farm machinery. In general, this resulted in the management of larger areas of land in much shorter time.

Signs of significant land degradation and of deterioration in water quality gradually appeared; there is now a realisation of how this has come about and of the extent of the damage (Allison and Martin 1994; Karssies and East 1997). Soil erosion was recognised in the 1940s and remedial measures are now incorporated into farm practice. Although this has not eliminated soil erosion, it is no longer the serious problem that it was. The most noticeable current effect has been the spread of secondary salinity resulting from inappropriate clearing and land management, resulting in rising water tables (Allison and Martin 1994, Holmes and Talsma 1981, Mulcahy 1983). Restrictions on clearing and a range of remedial measures, such as planting trees and other deep-rooted perennials, are now being practised. A more difficult problem to overcome is soil acidification (Co-operative Research Centre for Legumes in Mediterranean Agriculture 1996 pp. 50–51; Helyar *et al.* 1990; Karssies and East 1997). This arises under improved legume-based pastures or where some legume crops are used intensively in rotations. It occurs on poorly buffered soils and is more troublesome on sandy soils. To maintain the sustainability of the natural resources, farmers need to address these environmental issues, even though it may impose additional burdens on them.

With increased use of fertilisers and chemicals there has been a deterioration in the quality of water bodies and courses, adjoining agricultural lands. This is leading to practices of more careful fertiliser applications and more directed use of chemicals in farm practice. Effects on water quality have led to many studies under the auspices of environmental agencies.

With heavy machinery capable of covering large areas of land in a short time, compaction and loss of soil structure has become a problem. This has led to the practices of stubble management and incorporation and, more significantly, the use of techniques of minimum tillage. Minimum tillage has been rendered possible by the development of a range of chemicals and equipment to apply them. It has proved beneficial in a variety of ways. More recently, and in the course of development, is a wide tracked, multipurpose machine capable of cultivating, spraying, sowing and harvesting. The intention is to minimise the area compacted by traffic, using one set of tracks in the field for all operations. Stewart (1972) had considered a similar idea, proposed several layouts, and examined some technical issues. A development along these lines is a step towards driverless tractors, an idea that has been discussed intermittently, in the last 30 years.

Regular and continued applications of chemicals to control weeds has led to observations of herbicide resistance among some widely dispersed and persistent weeds, and this has become an acknowledged problem.

Remedial measures, some of which can be expensive, may affect the economic viability, and even the survival, of some farms. A diversity of Landcare programs initiated in the late 1980s has aroused awareness in the farming community of the need to modify management to contain the damage due to inappropriate practices, and many primary produces now participate in Landcare programs (eg McInnes 1995).

Hazards and risks

Although the words 'hazard' and 'risk' are regarded as synonymous in the lexicon, they are also used to differentiate between the influence of environmental and managerial factors.

Hazards such as frosts, drought, storms, cyclones and floods, pest and disease infestations and several forms of environmental degradation affect farming activities. There is no defence against cyclones or hail and little against extended waterlogging. While it is not possible to avoid such natural hazards, there may be mitigating techniques, depending on circumstances. For example some measures are available to reduce frost damage in intensive horticultural crops.

Plant breeders have developed rust resistance in wheat and continue to develop new lines as rust organisms mutate. This is true of many plant diseases. Lines that can withstand dry spells are in the very early stages of development and biotechnology may ultimately be successful in incorporating other physiological characters. In the meantime, cultural techniques for cropping in semi-arid environments have been developed to store water from early rainfall in the soil, for later use by crops at the crucial grain filling stage. Gradually, a variety of technological improvements are helping to overcome a range of hazards.

Risk is an unavoidable part of management and may be associated with financial management or operational decisions on the farm. Operational decisions embrace a wide range of activities such as choice of crop and/or cultivar, planting times, opportunity planting in the event of out-of-season rain, embarking on a new practice or system without prior trial, inadvertently bringing diseased animals onto the farm, etc. When making operational decisions farmers need to be aware of the consequences of risk, especially with respect to the occurence of hazards. These are often untimely and impose an unexpected financial burden, and farmers need to make allowance for them when developing overall risk management strategies.

White (1994) has categorised risks in farming as production, environmental, financial and marketing.

> 'Production risks cover those risks imposed by seasonal variability and unpredictability, certain manage-ment inputs, and the impacts of pests and diseases. Environmental risks include those associated with the degrading of soils, water, flora and fauna, and can often be reduced with proper management. These strategies should be combined with financial and marketing strategies in an integrated risk management strategy.' (McLennan 1996, p. 132)

Some risk may be intentional and some may be unwitting. Decisions involving risk can result in disastrous failures or spectacular successes and anything in between. Calculated risks, eg a calculated short cut, can result in a successful short-term outcome that may have detrimental long-term consequences to farm resources.

Intensification and diversification

Intensification and diversification usually take place in response to economic pressures.

Intensification becomes possible when some specific plant or animal character may open up the possibility of adopting techniques or management practices that were not possible hitherto. Two plant examples from overseas are used to illustrate intensification, because they are so stark. At the International Rice Research Institute (IRRI) the growing period of traditional rice was shortened markedly and rendered insensitive to photoperiod, so that two or even three successive crops of rice could be grown in the course of one year, thereby doubling or even trebling production. Second, at the International Centre for Wheat and Maize Improvement (CIMMYT) single stalk corn was developed and when these lines were grown at high densities under high fertiliser regimes, significant increases in corn yield were obtained. An example from the animal industries is the selection of sheep for twinning capacity to increase lamb production. More obvious examples are intensive piggeries, battery-raised chickens and cattle feed-lots. Intensification is more common in horticulture than in broadacre farming and in specialised industries.

Intensification may introduce the risk of catastrophic events. One such event in the early 1970s was the sudden and rapid spread of corn blight in the USA, mainly because of the narrow genetic base of the prevailing hybrid corn varieties. The risk of disease outbreaks in enterprises such as intensive piggeries, battery-raised chickens or glasshouse crops requires extremely careful attention to hygiene.

Primary producers need to be able to recognise the early warning signs of the damaging effects of intensification. For example the computer-based pest and disease warning system EPIPRE (Zadoks 1984) enables farmers to monitor the arrival and development of pests or diseases on their farms and decide when it becomes necessary to take defensive measures. It is a co-operative activity that relies on farmers sending their field observations to a central location where comparisons are made of the incoming data with the known epidemiology of the organism. A similar approach was used in an endeavour to control the insect pest *Heliothis* in cotton in the Narrabri district of NSW; an integrated pest management (IPM) program was devised (SIRATAC — Hearn and da Roza 1985) and used for a time. Although it is superseded now, it is cited here as an example of an integrated approach in pest management where intensive farm practices are in vogue. With livestock grazing, the economic incentive for higher stocking rates may lead to undesirable effects on either native vegetation or improved pasture. In this respect the rangelands of Australia are especially vulnerable. Supplementary feeding or de-stocking may be necessary, or alternatively multiple land use systems comprising production, conservation and tourism need to be found.

Diversifying can take the form of extending the range of crops that can be grown in particular environments (eg canola in the 1970s and more recently native plants as horticultural crops), developing pulse crops to grow in rotation with cereals in semi-arid environments (eg lupins), or introducing new breeds of sheep to meet specific market requirements (eg Awassi sheep from the Middle East into Western Australia). These are examples of straightforward diversification based on prior preparatory research programs.

Diversification can be used to reduce the risk of serious pest or disease infestation (Trenbath 1975, 1977). Experimental results have shown that mixtures of crops, or mixtures of cultivars with different degrees of susceptibility or resistance may reduce the degree of disease or insect infestations in crops (Wolfe *et al.* 1984). Farmers need to develop the capacity for flexibility in managing their resources and this may call for imaginative solutions that may not be obvious at first.

Issues for individual primary producers

Sustainable agriculture

There are a myriad of issues confronting primary producers today and into the future. The catchcry in the late 1990s is 'sustainable agriculture' (Australian Academy of Technological Sciences and Engineering 1994). Provided there is agreement on our earlier definition of farming systems, it follows that agriculture must be self-sustaining. Some primary producers fail and drop out, others replace them; this has always been so. It follows that the success of the individual enterprise rests on how skilfully managers use and deploy their resources and how effectively they allocate their time. This applies even more today than at any other time in history. This is the essence of the 'farm management movement' of the 1960s and 1970s when farmers were under economic pressure and they employed advisors to analyse and assess their performance, on the basis of farm budgets.

The margin between success and failure will become narrower. In the coming decade the demands on individual primary producers and their families will be more intense and they need to be even more skilled and have an even sharper vision of the tasks before them. Farming with greater precision, eg seed and fertiliser placement, timing of seeding in relation to soil moisture, etc., needs to be developed into a fine art. Farm incomes can be seriously impaired by unforeseen environmental influences such as drought or floods, hail or frosts, serious pest or disease outbreaks, or by errors of judgement, such as a sudden change in cultivar without adequate prior information, or the mis-timing of a particular operation, or incorrect setting on equipment. These factors can seriously reduce yield and lead to unexpected farm failures. What provisions need to be made to meet such eventualities?

Income, income supplementation and allied issues

Primary producers are at the mercy of prevailing markets and do not always receive fair prices for their products; witness the declining wool prices in the 1970s through to the 1990s (these were predicted by Anderson 1964) and the fluctuations in grain prices in the 1980s. They have little control over these and need to be able to produce sufficient volume to enable them to obtain a return that provides them with a living. For these reasons many primary producers diversify or restrict the scale of their operations to what they can manage. They do not always succeed and many farm families who do not earn enough seek off-farm employment to supplement their incomes and so continue to work their properties.

'In 1992/3 31% of broadacre family farms earned income from off-farm wages compared with 23% in 1988/89. … In the five years to 1992/93 there was a 40% increase in the amount of off-farm income derived by broadacre family farms.' (McLennan 1996)

This trend is likely to continue. Alternatives to off-farm incomes are partnerships with entrepreneurs willing to finance new ventures such as vine growing or feed lotting, diversifying into tourism activities such as farm stays, or leasing part or all of the property. There are also indications that successful business people such as accountants, transport operators and stock and station agents are investing in farms, or contributing to on-farm operations.

During periods of high interest rates, another area of concern to farmers is 'carry-on finance', to cover essential operations such as planting and harvesting crops, shearing or undertaking urgent maintenance. Because climatic factors have such a strong influence on productivity, and because the income stream of farmers can be so variable, an additional system of financing that is flexible and permits variation in repayments would greatly assist individual farmers to overcome adverse times such as droughts, especially when these periods are prolonged.

What is research contributing to improve income and where is it leading in terms of farm practice? How much are skills training and education contributing to improved farm management? On the social side, where are the community services located that are needed to sustain a rural population and how accessible are they? We pose these questions because in the total scene they form part of farmers' concerns.

Research needs to be closely integrated with the goals of the farming community. Research needs to be conducted at all levels; basic research underpins applied research and paves the way for developing new practices. There needs to be a balance between levels of research. Continuity in research is important — it is not something that can be turned off and on again from time to time. Research can assist in developing the quality characteristics needed for successful marketing and thus improve income. The rapid protein testing for segregating hard wheats in NSW, objective testing to improve wool classing, and eye muscle measurements to grade carcasses at or near the point of slaughter are examples where research has supported marketing requirements to develop suitable and rapid grading techniques. Research of this type needs to be extended to assist in value-adding to primary products and devising new uses for existing farm products.

Biotechnology is a generic term. It is genetic engineering, based on an understanding of gene structure. It has led to new developments and opened up many avenues in plant and animal breeding and improvement (Woolhouse 1994). It has accelerated the incorporation of pest and disease resistance, herbicide resistance, some physiological characters such as cold or waterlogging tolerance, and the development of improved quality characteristics and uniformity of products. For example, Hamblin and Atkins (1995), working with lupins, reported the incorporation of herbicide resistance and of high methionine seed storage protein, in a recently developed cultivar. They proposed to bestow these characters on older, successful cultivars and foreshadowed the transfer of genes for other characters.

In some crops, biotechnology may be more effective than traditional plant breeding in introducing desired characters and, in the future, may be used more extensively in conjunction with plant breeding programs. The value of current conventional plant breeding programs is in identifying suitable parents and in testing and comparing progeny. It is also useful in concentrating genes of small effect. These are important factors in developing commercial cultivars.

By next century, biotechnology will be an indispensable part of food production systems and will be integrated with traditional breeding methods. The practice of biotechnology requires sophisticated science, expensive laboratory equipment and people with superior research and manipulative skills. Biotechnology will be in the hands of scientists and their work will need to be well grounded in agricultural practice. Such people may be employed by scientific bodies, government laboratories, private companies and multinational corporations. Material from these programs will be subject to the same issues of ownership as arise under plant variety rights (PVR) and the ethics associated with PVR legislation.

Agricultural practices

Agronomy, animal husbandry and mixed farming practices will continue to progress and there are likely to be many new developments (Robson 1994). For example, the range of legume species and their uses will extend significantly (Co-operative Research Centre for Legumes in Mediterranean Agriculture 1996). There will be new ways of determining fertiliser requirements and methods of application, and these will permit variable fertiliser dressings in accordance with need. There will be further advances in the methods and practice of biological control and integrated pest management (IPM). Pest resistance to chemicals, as observed in bacteria, insects, worms and weeds, presents a challenge to overcome. Farmers and scientists will be pooling their practical and scientific knowledge to devise methods to combat such resistance. There will be innovations in agroforestry, alley cropping and mixed cropping. There will be improvements in flock or herd management with respect to nutrition and reproduction, based on better understanding of animal behaviour.

In more favourable environments land use will become more intensive but because of the fragility of the environment in marginal lands, management in those areas will become more cautious and conservative.

Because profit margins will become narrower and the cost of machinery and equipment will continue to rise, organisations will come into being that evaluate equipment and machinery. The role the Kondinin Group has undertaken so far is likely to extend. Innovations in machinery will come from users, as at present.

Education and skills training needs within the agricultural industries are as great as in other industries, but because the numbers in agriculture are small and dispersed, there is a problem in delivery. With new communication technology the problem of delivery can be overcome, but it will be more expensive, especially because effective training requires hands-on experience in a variety of areas such as animal husbandry, agronomy and horticulture and disease recognition etc. In this respect, regional universities such as Charles Sturt University and the University of New England will fulfil an increasingly important role — not only in agricultural education but in rural education as well. New communication technology will enable students in widely dispersed centres to participate in a single course. This is happening in TAFE which operates satellite centres in outback areas such as Mt Magnet in Western Australia. At the university level, since 1995 a nationwide unit in wool science is being offered at the University of New England, the University of New South Wales (until the end of 1997), the University of Adelaide and the University of Western Australia using tele-lecturing equipment. Units run jointly by these institutions are given in one place and transmitted simultaneously to the others. These lectures and tutorials are interactive and allow all the students to participate (Cooperative Research Centre for Premium Wool Quality 1995). From all accounts it is an exhilarating experience for those that have been involved.

This type of conferencing will come to be used more widely, not only in academic circles, but also in industry seminars and forums, in the exchange of information and data, and in trouble-shooting when unexpected problems arise.

Services

As services from State agriculture and other government departments diminish, commercial technical and managerial services will develop and their level of competence will vary. Farmers will need to evaluate carefully what such people have to offer and whether they are receiving value for money. Many of these services will rely on secondary information and on computer models devised by people with different backgrounds of experience, and operating in different environments. A farmer's scenario may be tested on several models before tendering advice — much as weather forecasters do nowadays by running their data through about four models and deciding how to present their forecasts. Some primary producers may prefer to access these models themselves.

Banking services in rural areas are shrinking rapidly and this is of some concern because it tends to limit the access to the monetary resource. It is important for farmers to plan their finances when they are devising short- and long-term management programs. It is difficult to predict how this will be resolved eventually, but it is conceivable that in the future, financing, accounting and taxation services may be combined with technical and managerial services and offered by a single body and this may not necessarily be a bank. The integrity of the people delivering such services needs to be assured.

The issue of occupational health and safety of primary producers is an essential component of effective management. The decline of rural medicine is now a matter of national concern. Regular monitoring of the health of farm families, of the effects of using chemicals, storage of hazardous materials, workplace layout and the design of machinery and equipment become important issues. Back problems and muscle ailments can be chronic and impair activities day to day. Accidents do occur on farms — some fatal and some near-fatal — and the time interval between the time of the accident and appropriate medical attention seems to be lengthening.

Some observations and questions

Australian agriculture has survived in the face of steadily declining terms of trade over a long period and while some would argue that it can continue to do so, it is difficult to see how this trend can continue. Margins of return to individual farmers continue to narrow. One consequence is that the number of people in primary production has declined significantly over the last 25 years. Another is that an increasing proportion of landholders seek off-farm incomes to maintain their enterprises or that business enterprises are becoming involved in farming. In future there may be a lesser proportion of family farms and a rising proportion of 'commercial' farms. We have also noted that because of natural hazards or errors of judgement, farmers are prone to risks of failure. Either productive capacity must take a quantum leap upwards, or prices for our agricultural commodities must improve significantly — preferably both. Perhaps this is what Woolhouse (1994) meant by the statement at the end of his paper and quoted at the beginning of this one.

The story of agriculture in Australia has been one of a resilient industry, in the face of many adversities. Some of the difficult times and events have passed into history. Australian farmers have responded by using their ingenuity to develop new land, new machinery and equipment and they have readily embraced new technologies and farming practices. A marked change that has virtually gone unnoticed, and does not appear to have received much attention from sociologists, is the nature of the farm population. We perceive a distinct change in the type of person engaged in agriculture. As a group, farmers today have different outlooks, different needs and aspirations than farmers a generation ago. It is likely that the nature of the farm population will keep on changing.

It is self-evident that farming systems and agricultural practice will continue to evolve. The question is: will the changes take place fast enough?

There are indications that we are about to enter a new wave of farm consulting that could develop into a new industry to service farmers. While in the 1950s and 1960s farm consultants dealt mainly with budgets and accounts, in the foreseeable future consulting will be concerned principally with technical and technological issues and the evaluation of the wide range of information concerned with farming.

We fervently believe that farmers and scientists must work closely together, to reap the full benefit of future advances in science and technology — not only biotechnology but other fields as well. Research has contributed significantly to maintaining and improving production. There is also a case to be made for keeping abreast with innovations in product development, information technology, marketing strategies and business practices. As profit margins keep narrowing and markets impose more quality demands, and the community imposes more stringent environmental requirements, it becomes *crucially* important to maintain the nexus between research and farm practice. The emphasis may move away from volume of production to quality, and also uniformity of quality. This means conforming to the strict product specifications set by markets. In our view, research needs to focus on both production and quality and needs to strike a balance between economic and ecological imperatives.

We should consider our food and fibre requirements in Australia, make the most efficient use of our most suitable agricultural lands and husband carefully our semi-arid and marginal lands. Earlier reviews of land use, such as that of Hallsworth *et al.* (1976), should be brought up to date from time to time. We must maintain our commitments as an international trading nation and strive to extend our export opportunities (eg Pasqual and Taylor 1995).

Perhaps we can pay more attention to value-adding to our agricultural products and also develop centres of production in different agroecological regions. An example is a small enterprise on the banks of the Murrumbidgee River near Jugiong, NSW, where asparagus is being grown, packed and exported. This activity sustains a small local community. If such small industries were repeated many times across regional Australia it would give a local focus, diversify our agriculture and contribute nationally.

There is a trend towards vertical integration. Many of our food processing industries seem to have become centralised in the major population centres and many are in foreign ownership (Vandore 1997). Does it need to remain that way?

It is generally acknowledged that in the developing areas of the world, expanding populations have a significant effect on the environment, on agriculture, and on agricultural practices. In Australia, we are not subject to the same severe population pressures, but we have a limit to the amount of land available for agriculture. Diminishing land area for agricultural pursuits and a rising population may combine to have deleterious effects on our fragile environment. Agriculture must avoid building up an 'ecological debt'. We need to be vigilant in monitoring population and environment and to be sensitive to changing circumstances and, in response, remain flexible in developing agricultural practices. As a society, we are good at quantifying but not so good at managing environmental resources. How successful are we at formulating policies concerning population, environment and agriculture (see Gifford *et al.* 1975) and what is being done to create opportunities for people?

Questions we might ask are: What policies do we have concerning population? ... food supplies? ... and preserving the environment? ... For example, what is an optimum population for Australia and for its various agroecological regions? ... Should we attempt to exploit the productive capacity of the different agroecological regions? ... Should we provide incentives? ... Should we question the accepted wisdom regarding 'economies of scale'? ... Tempting as it may be, it would be presumptuous of us to provide answers to such questions here.

Conclusion

We have endeavoured to portray the significant technical and technological developments in farming in recent times and to indicate that more are on the way. We have shown that agriculture is closely integrated with society and that as society changes, so will farming.

It is already evident that in order to conserve the natural resources of rural Australia, farmers need to adjust farming systems so as not to incur future 'ecological debts'. As well as responding to the community's environmental concerns, they have to contend with economic pressures and a changing social fabric. Cohesiveness within the farming sector may grow with increasing community demands.

If as a nation we are to become globally competitive, we would need to become internally efficient. This is a challenge to think about. There is neither a correct nor a single answer, and there are no simple solutions. It is only by discussions in various forums that a clearer picture will emerge of future directions for the farming environment.

Specific issues will need to be more clearly defined as we progress into the twenty-first century. We need to take the long view, think creatively and delineate the boundary conditions under which we operate, so that the farming community can share in the good things that are said to lie ahead.

Acknowledgements

W.J. Collins, J. Hamblin, D.R. Lindsay, R.J. Moir, D.J. Pannell and E.C. Wolfe made helpful comments on the manuscript.

References

Allison, G.B. and Martin, P.G. (1994). Soil, water and salinity — bringing science and economics together. In *Farming forever: Technologies for better crop production.* (Australian Academy of Technological Sciences and Engineering.) Invitation symposium, Melbourne. pp. 45–60.

Anderson, R. (1964). *On the sheep's back.* (Sun Books: Melbourne.)

Australian Academy of Technological Sciences and Engineering (1994). *Farming forever: Technologies for better crop production.* Invitation symposium, Melbourne.

Co-operative Research Centre for Legumes in Mediterranean Agriculture (1996). *Annual Report 1995–1996*. (University of Western Australia: Perth).

Co-operative Research Centre for Premium Wool Quality (1995). Wool is in the air. *Wool Press* **2**(1), 5.

Davidson, B.R. (1981). *European farming in Australia: an economic history of Australian farming*. (Elsevier: Amsterdam.)

Donald, C.M. (1982). Innovation in Australian agriculture. In *Agriculture in the Australian economy*. (Ed. D.B. Williams.) pp. 55–82. (Sydney University Press: Sydney.)

Donald, C.M., and Prescott, J.A. (1975). Trace elements in Australian crop and pasture production, 1924–1974. In *Trace elements in soil–plant–animal systems*. (Eds D.J.D. Nicholas and A.R. Egan.) pp. 7–37. (Academic Press: Melbourne.)

Fitzpatrick, E.A. (1964). Seasonal distribution of rainfall in Australia analysed by Fourier methods. *Archiv fur Meteorologie und Bioklimatologie*, Ser. B **13**, pp. 270–286.

Fitzpatrick, E.A. (1970). The likelihood of drought years in south-western Australia. *Journal of Department of Agriculture of Western Australia* (4th series), pp. 11, 144–146 and 155–164.

French, R.J., and Schultz, J.E. (1984). Water use efficiency of wheat in a mediterranean-type environment. 1. The relation between yield, water use and climate. *Australian Journal of Agricultural Research* **35**, 743-764.

Gifford, R.M., Kalma, J.D., Aston, A.R., and Millington, R.J. (1975). Biophysical constraints in Australian food production: Implications for population policy. *Search* **6**, 212–223.

Hallsworth, E.G., Martin, A.E., Millington, R.J., and Perry, R.A. (1976). *Principles of a balanced land-use policy for Australia*. (Land Resources Laboratories, CSIRO: Melbourne.)

Hamblin, A., and Kyneur, G. (1993). *Trends in wheat yields and soil fertility in Australia*. (Bureau of Resource Sciences, AGPS: Canberra.)

Hamblin, J., and Atkins, C. (1995). Genetically modified lupins: past, present and future. In *Herbicide-resistant crops and pastures in Australian farming systems*. (Eds G.D. McLean and G. Evans). pp. 35–40. (Bureau of Resource Sciences: Canberra.)

Hearn, A.B., and da Roza, G.D. (1985). A simple model for crop management applications for cotton (*Gossypium hirsutum*). *Field Crops Research* **12**, 49–70.

Helyar, K.R., Cregan P.D., and Godyn, D.L. (1990). Soil acidity in New South Wales — current pH values and estimates of acidification rates. *Australian Journal of Soil Research* **28**, 523–537.

Holmes, J.W., and Talsma, T. (Eds) (1981). *Land and stream salinity*. (Elsevier: Amsterdam.)

Karssies, L., and East, J. (1997). Australia's cereal producing lands: the state of the resource base. *Agricultural Science (New series)* **10**, 33–40.

Larson, W.E., and Pierce, F.J. (1994). The dynamics of soil quality as a measure of sustainable management. In *Defining soil quality for a sustainable environment*. (Eds J.W. Doran *et al*.) Soil Science Society of America, SSSA Special publication **35**, 37–51.

McInnes, K. (1995). Cooperative Landcare venture revisited. *Journal of Agriculture — Western Australia* **36**, 75–81.

McLennan, W. (1996). *Australian agriculture and the environment*. (Australian Bureau of Statistics: Canberra.)

Mulcahy, M.J. (1983). Learning to live with salinity. *Journal of the Australian Institute of Agricultural Science* **49**, 11–16.

Pasqual G., and Taylor, R. (1995). Food into Asia. *Journal of Agriculture — Western Australia* **36**, 94–100.

Perry, M.W., and D'Antuono, M.F. (1989). Yield improvement and associated characteristics of some Australian spring wheat cultivars introduced between 1860 and 1982. *Australian Journal of Agricultural Research* **40**, 457–472.

Robson, A.D. (1994). Impact of changing farming systems. In *Farming forever: Technologies for better crop production*. (Australian Academy of Technological Sciences and Engineering.) Invitation symposium, Melbourne. pp. 73–79.

Russell, J.S. (1973). Yield trends of different crops in different areas and reflections on the sources of crop yield improvement in the Australian environment. *Journal of the Australian Institute of Agricultural Science* **39**, 156–166.

Stephens, C.G., and Donald, C.M. (1958). Australian soils and their responses to fertilisers. *Advances in Agronomy* **10**, 167–256.

Stern, W.R. (1979). Trends and prospects. In *Agriculture in Western Australia 1829–1979*. (Ed. G.H. Burvill). pp. 380–397. (University of Western Australia Press: Perth.)

Stewart, G.A. (1972). Some new proposals for mechanization of cereal crop production in the tropics. *Journal of the Australian Institute of Agricultural Science* **38**, 29–36.

Trenbath, B.R. (1975). Diversify or be damned ? *The Ecologist* **5**, 576–583.

Trenbath, B.R. (1977). Interactions among diverse hosts and diverse parasites. *Annals of the New York Academy of Sciences* **287**, 124–150.

Vandore, J. (1997). Agri-food marketing — where now? *Agricultural Science (New series)* **10**, 31–32.

Walker, J., and Reuter D.J. (Eds) (1996). *Indicators of catchment health: a technical perspective*. (CSIRO: Melbourne.)

White, D.H. (1994). Managing risk on farms — living with uncertainty. *Resource Sciences Interface* **3**, 28–33.

Wolfe, M.S., Minchin, P.N., and Barrett J.A. (1984). Some aspects of the development of heterogeneous cropping. In *Cereal production*. (Ed. E.J. Gallagher.) pp. 95–104. (Royal Dublin Society/Butterworths:).

Wonder, B and Fisher, B. (1990). Agriculture in the economy. In *Agriculture in the Australian economy*. (Ed. D.B. Williams.) pp. 50-67. (Sydney University Press: Sydney.)

Woolhouse, H.W. (1994). Future direction. In *Farming forever: Technologies for better crop production*. (Australian Academy of Technological Sciences and Engineering.) Invitation Symposium, Melbourne. pp. 29-41.

Zadoks, J.C. (1984). EPIPRE, a computer based scheme for pest and disease control in wheat. In *Cereal production*. (Ed. E.J. Gallagher.) pp. 215–225. (Royal Dublin Society/Butterworths: London.)

Water, politics and power: can we integrate natural resource management in rural Australia?

Peter Cullen

Introduction

Natural resource management is about integration. It is about integration of the biophysical system, rather than treating land, vegetation, soil and water as isolated and separate sub-systems. It is also about integrating the biophysical and the social, including our beliefs about land and our expectations. It is about integrating our aspirations with the realities of world markets, where Australians are generally price takers rather than price setters.

Many rural and urban Australians are unhappy with the 'state of the bush'. They see television images of farmers undergoing great hardship, land being degraded and rural towns contracting. They see less of the success stories — of communities working together to solve problems, of entrepreneurial farmers developing new produce for boutique markets and of the general efficiency of much of our rural enterprise. It seems that most Australians would like us to achieve economic and ecologic sustainability in rural Australia and yet we seem a long way from achieving either of these goals. The Federal State of the Environment Advisory Council Report (1996) showed the loss of biodiversity and the degradation of land because of unsustainable land management that is causing severe degradation of our waterways, and in some cases these rivers are degrading coastal waters.

Water enters rivers either through groundwater flow or from surface runoff. Surface runoff in particular collects materials like soil particles, organic matter and nutrients and transports these to the river. Consequently we now appreciate that land and water are tightly coupled. Poor condition of a water body indicates poor land management in the catchment, and requires an examination of the various point and non-point sources of contaminants to the water and the various extractions of water which may be reducing the natural assimilation capacity of the river.

Ignorance will always be a problem with natural resource management. We are improving our knowledge base, but only through learning from the mistakes of the past. We do not yet have good models that let us predict land and water condition at the landscape scale. Science is an important part of land and water management in Australia, but there is still much to be done.

Table 3.1 The main symptoms and primary causes of degradation	
Symptom	*Primary causes*
Increasing stream salinity	Excessive clearing in upper catchments Excessive irrigation
Increased incidence of algal blooms	Excess nutrients from point and non-point sources Excess removal of water from rivers
Loss of biodiversity including fish	Dams and weirs blocking fish migration Trapping minor floods removes signals to biota Stream 'improvement' removes habitat
Loss of floodplains and riparian areas	Clearing for agriculture Levee banks prevent river and floodplain connecting
Invasions by pest organisms	Human stupidity with introductions Water management may optimise conditions for pests Inadequate rapid response capacity
Pollution and health risks	Sewage discharges from towns Agricultural wastes Agricultural chemicals
Excessive sand deposits	Catchment erosion

Knowledge has been important in documenting the environmental degradation of our water management, and in particular in promoting a 'whole system perspective' which includes the sites of downstream degradation in any analysis.

The main symptoms of degradation, and some primary causes are shown in Table 3.1.

The drivers of change

If we look at rural Australia we can see a number of factors presently shaping the future that appear likely to continue or accelerate.

Social

- *Population.* Rural populations and job opportunity will continue to decline except in rural–urban fringe areas. The rural population will continue to age as young people move to cities. Nationally the population will increase in the absence of any population policy and so will density as urban agglomerations grow and governments continue to avoid serious regional development policies.

- *Health concerns.* Our rural and urban communities are likely to demand higher standards for water and food to minimise health risks. These concerns will encompass use of growth stimulants and antibiotics in agriculture, and chemicals in land management, and may include attempts to control insect vectors by treating wetlands. Community concerns and market resistance to genetically altered produce is also a concern.

- *Ecologically sustainable development.* It is likely that community support for this notion will grow and governments will respond to it. International pressures will probably increase. There is a fundamental tension between agriculture, which seeks to maximise production and reduce variability through simplifying ecosystems, and ecologic sustainability which seeks to maintain diversity so that systems can cope with unusual events that may be catastrophic in a monoculture situation.

- *Community ownership.* This is the most exciting development of the last decade with communities realising that governments are not able to deliver a sustainable future and that communities are best placed to make choices and to integrate the various ideas being peddled to them (Cullen 1996).

Economic

- *Globalisation* of markets will continue to provide both opportunities and competition to rural producers. The demand for agricultural produce will increase with population growth. Whether it can be supplied at an affordable cost is uncertain. There will be ongoing pressures on the costs of production and non-tariff barriers to trade such as residues. Pests and disease risks will increase.

- There will be a continuing *retreat by governments* from investment in infrastructure development and renewal, and an expectation that private capital will provide these services. Since private capital will be seeking a profit, the costs of such infrastructure will rise, and producers will have limited bargaining power.

- The *economic rationalist* model will continue to demand short-term pay-offs and discount future degradation, thus skewing public and private investment.

- The use of *price signals* to lever change in natural resource management will increase. Pricing will be seen as a key tool in allocating scarce resources like water to the most efficient economic uses. This will have serious consequences for rural communities and individuals as they adjust.

Technological

- Much of the existing *infrastructure* is ageing and in need of refurbishment. This will provide opportunities to adopt new technologies rather than just replace with more of the same.

- There will be increasingly stringent *discharge rules* for farms as environmental regulators come to grips with non-point contamination of nutrients and agricultural chemicals, and continuing issues with residues in agricultural products. Approaches requiring 'best management practice' may stifle innovation.

- Rapid *advances* in remote sensing and real-time data capture and processing will make data cheaper and more timely, and will require better mechanisms to ensure the data get to decision-makers in a useful form.

- There will be continuing *innovation* in assessing crop water needs and timing, and the development of delivery systems that provide water of appropriate quality with less wastage. Improvements in system-level management are also likely.

Environmental

- There is improving *knowledge* of aquatic ecosystems and our predictive capacity as to how changes in land and water management will affect aquatic ecosystems. Better knowledge will allow better assessment of environmental allocations and how they should be used.

- *Rising groundwater* throughout the irrigation areas means that many farmers will be managing land with saline ground waters within a metre of the surface. There will be increased problems of discharge of saline drainage waters and possible increase in river salinity.

- Increasing *dryland salinity* along the western slopes will lead to loss of productive land and increasing salt loads to rivers.

- *Loss of biodiversity* in systems leads to a loss of resilience. Ecosystems will be less able to respond to unexpected environmental and climate changes because of this simplification (Cullen and Lake 1995).

- There will continue to be new *agricultural chemicals* released into the environment, and they and their breakdown products will continue to have unexpected and at time alarming effects. Considered responses will be difficult, given public concerns and our lack of knowledge.

The challenges to sustainability

There are a number of factors already evident in rural Australia that will continue to be barriers to effective natural resource management.

Greed

Greed continues to be a serious feature of natural resource management in Australia. Proposals to develop new water resources, such as those in Coopers Creek, with absolutely no attempt to understand the ecological consequences, will continue, and governments will be pressured to bow to these 'developments'. Governments are slowly learning that it is taxpayers that pick up many of the costs of these poor decisions, and the taxpayers are starting to recognise that the beneficiaries of these ill-considered schemes rarely pay the costs.

Greed needs to be exposed, especially when private profit is gained by passing on the real costs of production to downstream water users and the environment. There are some who might argue that as price takers some farmers have no choice but to pass on these costs to the community. If this is the situation, then presumably members of the community have some say as to whether they wish to be involved in such enterprises, and so we may need to have a stronger system of rural land use planning. Certainly these are community issues which need effective mechanisms to reconcile the various interests concerned.

Ignorance

Much of what the early explorers saw was the reverse of their expectations. Mountains had rounded tops unlike the scraggy rocky cliffs of Europe. Rivers did not conform to their expectations of large rivers draining the country and providing navigation from the sea but, as Sturt recorded, '… they fail before they reach the coast and exhaust themselves in marshes'.

Cook's description of Botany Bay and its surrounds as some of the finest meadows in the world excited people in England, leading to the First Fleet. Yet within 6 months of arrival, Tench, quoted by Carter (1987), wrote 'Of which the natural meadows which Mr Cook mentions near Botany bay, we can give no account.' A year later Tench wrote (September 1789):

> 'Had not the nautical part of Mr Cook's description been so accurately laid down, there would exist the utmost reason to believe that those who described the contiguous country, had never seen it.'

Seasonal variability was something the newcomers could expect, but the annual variability in Australian rainfall was outside their experiences. This misunderstanding of the Australian climate led to a plethora of poor decisions of land settlement made with no understanding of land capability (Cullen and Lake 1995). These were not just early mistakes, but continue to the present.

Institutional arrangements

The boundaries that form our States can now be seen to be historical accidents that have impeded the development of our country. Using a river like the Murray as a State border is about the worst possibility, given our present understanding of the need to manage on a catchment basis. There is little likelihood of getting more sensible boundaries for regional government.

The Murray-Darling Basin Initiative is an internationally recognised example of managing a water resource across jurisdictional boundaries. But it has a limited scope, and little power to influence States who seek to repeat mistakes already made by others. The rapid changes in the water and land management organisations mean that the leaders who developed the initiative have now been replaced by a new breed of manager who have yet to demonstrate the vision or commitment to land management demonstrated by their predecessors. Hopefully they have a similar vision to integrated resource management rather that short-term bottom lines and maximising returns to their State. The jury is still out.

Fashions in government

The current fashions of government for downsizing, privatising, competition and outsourcing are not likely to lead to more effective or even cheaper public management of natural resources, based on experiences so far. They will lead to a narrower focus, pressures to divest agencies of public good obligations and a short-term view. They will require more and smarter regulation. They have already led to a dramatic shedding of intellectual capacity in the public agencies responsible for resource management. Appropriate pricing of public assets is a considerable challenge. Where governments seek a quick sale they may underprice the assets.

The emerging paradigm

There is a widespread acceptance that the nineties are seeing the emergence of a new paradigm for water management, although there is less agreement as to what it might contain. I expect the following features to be important elements of the emerging paradigm:

- There will be a better appreciation of all the sources, stores and sinks for water and contaminants as they enter and move through the landscape. This will be done on a catchment basis, and will come to be viewed as rate of movement problems rather than mass stored problems.

- Sustainable development means identifying the key ecological processes that determine the condition of waterways, and ensuring that these processes are maintained sufficiently to provide an acceptable environmental outcome.

- Water needs to be of an appropriate quality for the use proposed, rather than demanding high quality drinking water for all uses. This will involve planning processes to allocate water to uses and bargaining to select appropriate water quality guidelines for uses in particular catchments.

- Pricing and public education will be key ways that change community perceptions of water and the demands on water resources for both the extraction of water and for the disposal of water. The community will be more involved in setting standards, assessing risks and understanding the trade-offs required.

- It is clear that our present strategy of splitting the water cycle into a whole range of different elements, managed by different groups of professionals, has now outlived its usefulness and we now need to put the pieces of the jigsaw back together again.

- The aquatic ecosystem should be what drives the technologies used, and so ecologists need to be involved with planners and engineers to develop more integrated solutions and set appropriate performance criteria for urban stormwater and waste treatment.

- Agency structures need to reflect this need for biophysical integration through a variety of mechanisms including integration and co-ordination.

Regional water quality management requires a consideration of point and non-point inputs to waterways, and a consideration of the likely change these inputs will have on the receiving ecosystem. Such planning should be undertaken for high, medium and low flow conditions. The impacts of sewage discharge and of inadequately treated wastes from intensive animal industries are critical inputs to catchment-based water quality planning.

It is becoming obvious that the best way forward is to make our decisions within a whole system context. There is little point capping water extraction from the Murray-Darling if many people can just start exploiting groundwater. I think it is silly to develop salinity and drainage strategies for the irrigation areas without thinking about the sources of salt from the dry-land upper regions of the catchment.

The total catchment management model is now widely understood in Australia. The difficulties with using it, however, include:

- when upstream people are a long way from and have little contact with downstream people, especially when they are in different industries (eg tourism and fisheries compared to conventional agricultural industry);

- when some impacts may take a long time to appear, and may be dispersed, affecting a lot of people to a small extent, whereas the beneficiaries may be a small number who benefit greatly (the 'tragedy of the commons' argument).

Challenges for the political and economic system

I have an ecologist's view of sustainability. I see five significant problems outside ecology that must be addressed.

1 Most economic analysis and decision making is very short term, yet the problems of ecosystem management and land and water degradation may well be in the 10–50 year time scale. The economist's use of discounting seems to make such degradation almost without cost.

2 Economic thinking appreciates the importance of externalities, but appears not to have appropriate tools to consider them. They are a critical element in land management.

3   The ecosystem is unable to differentiate between private and public lands, and yet our organisational frameworks have great trouble getting appropriate cost sharing principles where public expenditure on private land may provide significant public benefits. The Community Advisory Committee of the Murray-Darling Basin Managerial Council (MDBMC) has done important work on cost-sharing principles, which governments have yet to respond to. I appreciate the problems of public funds again being translated into capital land values for private gain. The idea of capturing these capital benefits on sale of the property is worthy of exploration, based on the delayed payment model used by HECS in education.

4   When the catchment crosses State boundaries, and while commonly the local residents understand and wish to work on a whole system basis, the State and Federal agencies are unable to cope effectively with this.

5   Institutional arrangements are changing frequently in land and water management, with little comparative analysis or evaluation. The 'reforms' often lead to a loss from the organisations of technical capacity, which is being outsourced or even dispensed with. The past institutional arrangements are probably a key cause of much of the present degradation. Our organisational arrangements for managing natural resources in any integrated way have failed. Our professional training, professional organisations and agency structures all evolved in simpler times. The pressures on resources were less then and we had only simplistic understandings of the cause-effect relationships in ecosystems. It is little wonder they are not coping with the challenges of the nineties and beyond. There are tensions between the various agencies and the interests they reflect, and integration has generally been poor. The narrow sectorial interests tend to discount externalities such as downstream degradation and any impacts that will not be apparent for some time.

Irrigation futures

Irrigation gives us a useful case study to see how the various drivers identified are being applied to a particular industry. Irrigation is going through a revolution, and the Murray-Darling Basin provides a good case study. The Ministerial Council decided to cap extraction from the Murray-Darling Basin. Given that the river was showing extreme signs of stress with toxic algal blooms, and that around 80% of the median flow was being extracted from the river, this seemed one way of at least preventing the problems getting worse (Cullen *et al.* 1996). The independent audit process has produced a new level of accountability and public exposure as to who gets water for what purposes.

The Council of Australian Governments (COAG) processes provide strong financial incentives for States to conform to agreed principles. The acceptance of full-cost pricing will change the face of irrigation in the basin. This will introduce greater efficiency of water use, and encourage water to be used on high value crops. Whole industries are likely to decline under this new economic regime, but other higher value crops will be able to get water to expand. The clarification of water rights based on a separation from land rights will provide a basis for water trading which will accelerate this readjustment. The formal provision of a water allocation for the environment based on the best available scientific advice is also required.

In a recent US study on futures for irrigation, the National Research Council (1996) identified some future trends for irrigation in the US which look quite likely also to be experienced in Australia:

• Area irrigated will decline but the value of production will stay the same or increase;

- Amount of water supplied to irrigation will decline;

- Economics of irrigated agriculture will be driven by global markets;

- There will be a shift to large, well funded operations;

- There will be increased environmental regulation; and

- There will be improved efficiency due to advances in irrigation technology.

Some ways forward

We need to appreciate that all of our interventions in a catchment are driven by ecological imperatives that we need to try and understand. We have been slow to appreciate this, since ecological impacts such as salinisation may not become apparent for 50 years after we have removed vegetation or over-irrigated for many years. Humans have trouble understanding impacts at these time scales, and our predictive tools are still weak, although they are improving.

Proposition A — Ecology sets what is possible

We are dealing with ecological systems. The ecosystem itself provides the design criteria for our interventions, not what is technically or economically easy or feasible. The appropriate time and space scales are those of the ecosystem rather than human preferences.

The power of Landcare and Total Catchment Management has been in local people collecting and understanding information relevant to their land and their problems. Agricultural extension services have largely collapsed and been partially replaced by Landcare facilitators who bring information to the group and allow the group to decide if the knowledge is useful to them. The knowledge is continually being tested against the reality as understood by the landholders.

Proposition B — A learning community?

We do not know enough to be able to predict how ecosystems or institutional arrangements will respond to changing circumstances. All of our interventions need to be seen as experiments where we do what we think is the most appropriate thing, but then review and analyse the outcomes we produce. We need a continuing learning approach as we test our ideas against reality. We must build our capacity to be a learning community and welcome challenge and evaluation.

The Landcare model uses a whole variety of local pressures to take encourage compliance, and these pressures from neighbours cannot be ignored in the way an agency extension officer could be ignored. The key to this is to make sure we have mechanisms to provide technical information to these leaders and let them choose to demonstrate and push particular approaches. The agency staff are knowledge providers, not de facto managers of the land. They have to have time to build trust with key local influencers.

The knowledge providers from research have to re-examine how new findings can be got out to the farming community. No longer can agricultural experimental stations be relied upon to test and demonstrate new approaches. These demonstrations will be done with leading farmers. There remains a challenge of getting highly technical, disciplinary based information presented in a form that can be used effectively by community groups.

Proposition C — Integration is fundamental

Simplifying problems down to their sub-components has worked well for much of science, but does not provide the answers to large-scale ecological problems. We need to lift our skills in integrating the contributions from different disciplines, in thinking in a whole system context, and including health, economics and environment as parts of a solution, not isolated bits that can be sorted out separately.

Summary and conclusions

A sense of community is fundamental to us moving forward and tackling the difficult challenges facing rural Australia. In the absence of a set of 'right' answers that we can agree upon, it is essential that we build our capacity as a learning community. Science contributes much knowledge and many tools. But perhaps more importantly it contributes a framework of organised scepticism that lets us accept interim solutions, and then test them against the reality of the outcomes they produce. This culture of evaluation and review needs further development in Australian natural resource management.

References

Carter, P. (1987). *The road to Botany Bay: an essay in spatial history.* (Faber & Faber: London.)

Cullen, P. (1996). Empowering communities — challenges for technical communicators. In *Proceedings of Stockholm Water Symposium Safeguarding Water Resources for Tomorrow. New Solutions to Old Problems.* Publ. 6. Stockholm Water Symposium.

Cullen, P., Doolan, J., Harris, J., Humphries. P., Thoms, M., and Young, B. (1996). Environmental allocations — the ecological imperatives. In *Managing Australia's inland waters. Roles for science and technology.* (Prime Ministers Science and Engineering Council, September.) (Department of Industry, Science & Tourism: Canberra.)

Cullen, P., and Lake, P.S. (1995). Water resources and biodiversity: past, present and future problems and solutions. In *Conserving biodiversity. Threats and solutions.* (Eds R.A. Bradstock, T.D. Auld, D.A. Keith, R.T. Kingsford, D. Lunney, and D.P. Siverton). pp 115–125. (Surrey Beatty & Sons Ltd)

National Research Council (1996). *A new era for irrigation.* pp. 203. (National Academy Press: Washington.)

State of the Environment Advisory Council (1996). *Australia: state of the environment 1996.* (CSIRO Publishing: Melbourne.)

Challenges for conserving biodiversity in Australian freshwater ecosystems

Robyn J. Watts

Introduction

Worldwide there is increasing pressure on freshwater resources. For instance, in the nineties we have seen a worldwide increase in the number of dams constructed, with a shift towards larger dams (Gardner and Perry 1995). The same pressures on water resources are occurring in Australia, where our extremely variable hydrological cycle has meant that water storages are larger per head of human population than elsewhere in the world.

Until recently little attention has been given to conservation of biodiversity in freshwater systems relative to issues of conservation in Australian terrestrial environments (Cullen and Lake 1995). This is now changing, as water allocation and land activities that impact on freshwater systems are the focus of much public and research interest. However, documentation and conservation of biodiversity in Australian freshwaters remains patchy. Here I summarise biodiversity loss in Australian riverine systems and discuss the challenges for conserving biodiversity in Australian freshwaters.

What is biodiversity?

The term 'biodiversity' is usually used to represent the variety of life; however the term has great breadth. It can be regarded as a concept, a measurable entity, or a social/political construct (Gaston 1996).

Several schemes distinguish different levels of biodiversity (eg Noss 1990; McAllister 1991; Soule 1991), but it has become widespread practice to define biodiversity at three hierarchical levels: genetic diversity, species diversity and ecosystem diversity. The National Strategy for the Conservation of

Australia's Biological Diversity (Commonwealth of Australia 1996) and the Draft NSW Biodiversity Strategy (New South Wales National Parks and Wildlife Service 1997) recognise these three levels.

Each of these levels of diversity usually has three attributes: composition, structure and function (Noss 1990). Composition refers to the variety and identity of elements, such as the number of genes, the association of species or the distribution of habitat patch types. Structure refers to the physical organisation of a system, such as the number of layers in a macrophyte community or the distribution of age classes of a species. Function involves processes, such as nutrient cycling rates, recruitment rates or the rate of gene flow.

Thus, biodiversity includes all species and the genes they possess as well as the variety of habitats, communities and ecological processes that occur in systems. For freshwater ecosystems biodiversity incorporates both instream (the river) and floodplain components (lakes, overflows, wetlands and so on).

Biodiversity losses in freshwater systems

Two hundred years of European occupation has significantly altered freshwater and terrestrial ecosystems in Australia (Saunders *et al.* 1990). Biodiversity in freshwater ecosystems is especially vulnerable because human populations and most of our agricultural activities depend on freshwater and are concentrated near them. Furthermore, as energy and materials usually flow from terrestrial to aquatic systems, most activities in the terrestrial environment ultimately affect adjacent aquatic systems.

Loss of freshwater biodiversity is due to multiple disturbance factors. In Australia these factors include alteration to the hydrological cycle (Close 1990; Walker *et al.* 1995), direct and indirect alteration of habitat (eg Robertson 1997), pollution (eg Fairweather 1998), introduced species (eg Fletcher 1986; Harris 1997) and commercial and recreational exploitation of species (Wasson *et al.* 1996).

Even though the physical, chemical and biological degradation of freshwater environments is widely recognised as a major problem, the loss of diversity in freshwater systems has received comparatively little attention (Moyle and Leidy 1992). There is often anecdotal evidence for a reduction in the biodiversity of many freshwater ecosystems (eg Roberts and Sainty 1996), but limited quantitative evidence.

Fish are good indicators of trends in freshwater biodiversity because they occur over a wide range of habitats and play an important trophic role as both predators and prey. Thus they have a major impact on the distribution and abundance of other organisms. Most freshwater fish faunas around the world are in serious decline (Moyle and Leidy 1992). In North America 3 genera, 27 species and 13 subspecies of fish became extinct during the past century (Miller *et al.* 1989), and habitat loss and species introductions were considered to play a significant role in these species extinctions. As can be seen in Table 4.1, about one-third of Australian freshwater fish species are threatened or rare (Saunders *et al.* 1996). The increase in the numbers of threatened species of fishes in Australia has been accompanied by declines in population sizes of many species (Reynolds 1976; Harris and Gehrke 1997), suggesting that many populations are more fragmented than they were in the past. A good example to demonstrate this comes from a recent regional survey of fish in New South Wales. The data from this survey show that more than 99% of the fish in the Murrumbidgee

Table 4.1	The conservation status of Australia's freshwater fish.	
Conservation status	Number of species	Percentage of estimated total
Presumed extinct	0	0
Endangered	9	4.1
Vulnerable	8	3.7
Poorly known	19	8.8
Rare	36	16.6

Source: Saunders et al. 1996

catchment are carp, which are introduced noxious fish (Harris 1997). Many of the factors thought to be responsible for reduction in fish biodiversity include changes to river-flow regimes (eg channel flows and links to adjacent wetlands), direct alteration of habitat, introduced species, pollution and commercial and recreational exploitation of species (Cadwallader 1978; Moyle and Leidy 1992; Allan and Flecker 1993; Gehrke and Harris 1994; Noss and Cooperider 1994; Gehrke et al. 1995; Mallen-Cooper et al. 1996; Harris 1997).

Although the loss of diversity in native freshwater fishes and regional and local losses in waterbirds is reasonably well documented, we do not have as good an understanding about loss of species biodiversity for other taxa. There has been significant regional alteration to native vegetation on floodplains (Smith and Smith 1990), and regional and local losses of plants in rivers and wetlands (Roberts and Sainty 1996; Robertson 1997), both of which are likely to have influenced amphibian regional distribution and abundance (Healey et al. 1997). Damming of rivers and alterations to flow have significantly altered the local and regional distributions of crustaceans and gastropods in nearly all Australian catchments (Walker 1985; Horwitz 1994; Saunders et al. 1996). In some cases, eg freshwater gastropods, regional extinctions have been reported (Walker et al. 1992).

Information on the loss of genetic diversity is virtually non-existent. A few studies (eg Keenan et al. 1996) suggest that genetic diversity has been lost in some freshwater taxa, but evidence is very limited.

The extent to which ecosystem diversity has declined is reasonably well documented. Overall the greatest losses and degradation of wetlands have occurred in areas used for cropping, pasture, plantation and urban development (Wasson et al. 1996). There have also been severe reductions in habitat quality and connectivity between habitats as a result of river regulation and extractions of water (Walker 1985). There is increasing evidence that fundamental ecosystem processes in freshwaters have been altered to the extent that ecosystem function is being lost (eg Robertson et al. 1996).

Challenges for conserving biodiversity in Australian freshwaters

The reasons for conserving biodiversity are generally grouped into four categories: ecosystem processes (ecological reasons), economics (utilitarian reasons), ethics (moral reasons) and aesthetics (recreational reasons) (Saunders et al. 1996). Whilst people with different views often disagree on the relative importance of these four categories, there appears to be agreement on their existence. Regardless of our philosophical reasons for conserving biodiversity, the overriding aim is to maintain biodiversity through conservation and co-operative management and to provide sustainable development in rural and urban areas. I will briefly discuss the social, research, management and economic challenges confronting us if we are to achieve this goal.

Social challenges

Scientific knowledge and government policies alone will not solve the problem of biodiversity loss. We need a collective social responsibility if we are to conserve biodiversity in Australian freshwater ecosystems (Stafford-Smith *et al.* 1997).

The primary social challenge for us as a community is to assess the importance placed on the conservation of biodiversity in freshwater ecosystems. The NSW Minister for the Environment poses this question to readers in her introduction to the proposed interim environmental objectives for NSW Waters (Environment Protection Authority 1997): 'How healthy do you want our waterways to be?' Bowmer (this volume) discusses how myth, memory, art, advertising and the media shape our perceptions and expectations of the Australian landscape and how community perceptions have influenced policy development in this country.

The discussion of biodiversity in the community has focused on a few large animal and plant species but little on what the term 'biodiversity' really represents and how biodiversity is essential for the maintenance of ecological systems. Unless the general public has an informed view of biodiversity and why it is important to their lives it will be difficult to achieve the support for policies that will ensure the conservation of biodiversity. A study of the Landcare program suggested that there is some evidence that Landcare participants understand the link between sustainable agriculture and biodiversity conservation (Curtis 1997); however the study also recognised the need for clearly articulating the linkages between conservation of biodiversity and profitable agriculture. On the basis of this study Curtis (1997) suggested that Australian policy makers would be better advised to focus on awareness raising and landholder training than upon changing landholder attitudes.

Several recent forums have attempted to encourage dialogue between policy makers, scientific researchers, industry, conservation groups and interested members of the public (eg Pigram and Hooper 1992; Kingsford and Porter 1997). More recently, as part of the NSW water reforms the NSW government has formed river management committees in several inland river catchments. These committees are comprised of government representatives and community members (eg landholders, irrigator groups, environmentalist groups, local government, Aboriginal organisations). The role of these committees is to review the environmental flow rules proposed by the government for their catchment and develop alternative flow rules that are agreeable to all groups. These forums have improved the dialogue between researchers, managers and the community. We need to have continued constructive dialogue that results in both environmental benefit and sustainable development.

Biodiversity conservation cannot be achieved solely by the creation of a protected area system and implementation of government policies as a large proportion of freshwater habitat in Australia is contained within freehold land or has freehold land adjacent to it and is thus affected by land management practices. For example, 90% of floodplain wetlands in the Murray-Darling Basin are privately owned (Murray-Darling Basin Commission 1995). The successful implementation of government policies on biodiversity conservation therefore relies on the co-operation of private landowners. Community partnerships such as Landcare and Rivercare were established to promote ecologically sustainable agriculture and enhance biodiversity (Farley and Toyne 1989). These groups have mobilised a large section of the rural population and have increased awareness of issues. Most importantly, these groups have accomplished projects that are likely to have an impact on land and water degradation at the local scale (Curtis and DeLacy 1995). We need to support these community networks and individual landholders so that they can implement management on private lands that will complement the conservation efforts on public lands. Support will be required on both training, information and economic fronts (Curtis 1997).

Research challenges

Our knowledge of biodiversity in freshwater ecosystems is continually improving, but there are still considerable knowledge gaps. We need to improve our knowledge of what biodiversity we have, what we are losing, and what we can do to prevent further loss of biodiversity.

To date there has been little focus on genetic diversity in freshwater systems as this area of research is generally afforded lower priority than other management concerns (World Conservation Monitoring Centre 1992). In contrast, our understanding of diversity at the species level is much better than that at the genetic level. The diversity of freshwater vertebrates and higher plants is generally well known, mainly because of their size and comparative ease of identification. The composition of macroinvertebrates is becoming increasingly well documented; however, new taxa are still being identified. In contrast, our knowledge of the smaller freshwater invertebrates (eg plankton) and other micro-organisms (eg bacteria) is poor, even though they no doubt represent a large proportion of the species in freshwater systems and play an important role in freshwater foodwebs. We need to improve our knowledge of diversity for all attributes of composition, structure and function. This is particularly true for inland freshwater systems because a large proportion of the research in Australia has focused on freshwater systems close to our major cities. As there are limited resources for biodiversity studies, the challenge is to focus and co-ordinate new research so that we fill the knowledge gaps most efficiently.

As the theory of reserves is based on terrestrial models, we need to develop better models specifically designed for freshwater systems to ensure links between freshwater habitats are maintained. The existing network of freshwater reserves in Australia is derived from a variety of sources ranging from the Ramsar Convention (Australian Nature Conservation Agency 1996) to freshwater reserves established under State fisheries agencies (eg NSW Fisheries 1998). The Ramsar Convention identifies nationally important wetlands based on criteria representative of rare wetlands, rarity of species, and wetlands with special hydrological, ecological or cultural values. In contrast, the network of fisheries reserves is a system of Closed Waters and Notified Trout Waters that are set aside with the aim of protecting breeding populations of specific fish species (eg NSW Fisheries 1998). I believe that the network of freshwater reserves in Australia needs to be re-examined. Research is required to determine a more comprehensive approach to developing a reserves network that will maintain linear links within rivers and lateral links between the river and the floodplain. We also need to consider the role of freshwater habitats on private lands, as whole catchments cannot be protected and land practices upstream of reserves will impact on the habitat downstream. To date, most of the discussion and research regarding the inclusion of private lands in a protected area network has related to terrestrial systems (eg Binning 1997). The challenge is to extend this discussion into freshwater systems.

Management challenges

As a signatory to the international Convention on Biological Diversity Australia is committed to the conservation of biodiversity and the sustainable use of ecosystems, species and genetic resources. The Australian Commonwealth, State and Territory governments are all responsible for managing human activities that threaten biodiversity in their jurisdiction. There are several challenges for managers in both the development and implementation of policy related to the conservation of biodiversity in freshwater systems.

Until recently, water resource managers have primarily focused on the delivery of water for agriculture, urban and industrial purposes. The increase in community awareness of environmental issues has encouraged governments to consider the development of policies that embrace the concept of ecologically sustainable development. This has resulted in the development of policies which reflect a more balanced approach to resource use, one that gives consideration to environmental, social and economic issues (McPhail and Young 1992). Thus, water allocation for environmental purposes, such as wetlands replenishment, is now on the policy agenda of many levels of government. For example, the Murray-Darling Basin Commission adopted the cap on water allocations (Murray-Darling Basin Ministerial Council 1996) to attempt to address the problem of unsustainable growth in diversions of water within the Murray-Darling Basin. Similarly, the NSW Department of Land and Water Conservation has recently implemented their water reform process which includes an assessment of a range of catchment issues from water allocation to weir reviews (NSW Department of Land and Water Conservation 1997). The challenge for managers is to ensure that these policies can be implemented for an ecologically appropriate length of time, as they will not always result in immediate environmental benefits. There is also a need for scientists and scientific societies to become more involved in the development of policies and to ensure that their research findings are considered during policy development. Many scientific groups (eg Australian Society for Limnology) are currently discussing this challenge (Australian Society for Limnology 1998). Many scientists believe that their role is to provide knowledge without being involved in social or political issues, whereas others suggest that scientists have an important role to play in social and political debate and policy development.

Managers often face the problem of having to make decisions before scientific certainty can be given, which can lead to situations where managers, scientists and the community are in conflict. In Australia we have recently seen the development of national policies on issues such as ecologically sustainable development (Commonwealth of Australia 1992) and biological diversity (Commonwealth of Australia 1996). These types of policies place new demands on policy makers and managers (Dovers and Mobbs 1997). A few of the new demands identified by Dovers and Mobbs (1997) include the need to integrate social, environmental and ecological considerations in policy, increase research, monitoring and communication efforts, and adopt the precautionary principle without allowing scientific uncertainty to delay action when impacts may be serious or irreversible.

The adaptive management approach has been suggested as one of the ways management can move forward and meet these new demands. The aim of adaptive management is to develop policies so that hypotheses can be posed and tested, combining the rigours of scientific method and the realities of management. Thus, policies can effectively become a means of learning more about the system, rather than a indisputable procedure. This implies that there will be sufficient monitoring prior to and during the implementation of the policy to enable the results of the management change to be determined. Management decisions can then be refined and changed over time as better information comes to light. Managers, policy makers and politicians must allow new policies to be implemented for a sufficient length of time to allow them to be evaluated. If policies are changed before they can be assessed then the rigour of adaptive management will break down.

As discussed above, multiple factors have caused the breakdown of our freshwater ecosystems. Managers need to implement policies that will allow a co-ordinated approach to addressing these factors.

Integrated catchment management (ICM) has been suggested as the way forward for management of natural resources. ICM is the co-ordinated and sustainable use and management of land, water and vegetation and other natural resources on a water catchment basis so as to balance resource utilisation and conservation. ICM has the potential to play an important role in the conservation of biodiversity by encouraging links between the community, government, industry and science in the management of natural resources. As discussed by Booth and Burgin (1997) and Bellamy *et al.* (1996), the implementation of ICM is still in its infancy and there is much to achieve. Although there have been considerable developments in ICM, the evaluation of the effectiveness of the process in resolving land and water management problems in Australia has been neglected (Bellamy *et al.* 1996). The challenge for ICM is to support the on-ground community activities so they are truly integrated across land and water components of the catchment. Also there is a need to evaluate the progress and effectiveness of catchment management programs using a range of indicators, such as those developed for national state of the environment reporting (eg Fairweather and Napier 1998; Saunders *et al.* 1998).

Managers also face the challenge of addressing the expectations of the community, which often expects quick solutions to problems. Managers need to have a constructive dialogue with the community and clearly inform its members that there are no simple answers to the complex environmental issues that influence biodiversity. If the community is sold simplistic solutions that are unachievable, then its members are less likely to support the long-term conservation efforts that are required if we are to conserve biodiversity in freshwater ecosystems.

Economic challenges

Biodiversity conservation is as much about economics as it is about ecology. Research, monitoring programs, education programs and rehabilitation programs all require financial support. If the community is earnest about conservation of freshwater biodiversity then there must be some collective responsibility for the rehabilitation of catchments and the appropriate funds must be made available. Is the community prepared to pay for biodiversity conservation, even if the benefits may take a long time to materialise?

In difficult economic times the temptation of governments will be to cut costs on issues such as biodiversity conservation. As the loss of biodiversity is overlooked in the national accounts, there will not always be an economic argument to continue funding for biodiversity programs. It is important that the economic benefit that biodiversity brings to this country is considered when making decisions on funding for biodiversity conservation.

Funding for research and monitoring is as important as funding for rehabilitation programs. For example, adequate funds must be provided for the scientific assessment of management changes, to ensure that there is sufficient statistical power to determine the effect of a management change.

We also need to consider the cost of restoration of freshwater habitats on private land. Landholders cannot afford the full expense of restoration and conservation and there is a high level of anxiety that these changes will be imposed upon them (Curtis 1997). We need to decide if there is a role for financial incentives in achieving off-reserve conservation and how this could be best implemented (Binning 1997).

As a community we may also have to accept reduced profits today to conserve biodiversity and implement sustainable practices which will ensure the long-term ability of our lands and associated freshwater systems to provide profits in the future.

Many of these economic challenges return us to the social challenges which began this discussion. As a community we need to decide the importance we place on the conservation of biodiversity relative to other issues. To make this decision we need to be fully informed of the importance of biodiversity and the role it plays in maintaining and sustaining our world.

References

Australian Society for Limnology (1998). The future of ASL. *Australian Society for Limnology Newsletter* **36**(2), 36–41.

Australian Nature Conservation Agency (1996). *A directory of important wetlands in Australia.* (Australian Nature Conservation Agency: Canberra.)

Allan, J.D., and Flecker, A.S. (1993). Biodiversity conservation in running waters. *Bioscience* **43**, 32–43.

Bellamy, J.A., MacDonald, G.T., Syme, G., and Johnson, A.K.L. (1996). The implementation of Integrated Catchment Management: an environmental, social or institutional problem? In *Downstream effects of land use.* (Eds H.M. Hunter, A.G. Eyles, and G.E. Rayment). pp. 403–404. (Department of Natural Resources: Brisbane.)

Binning, C.E. (1997). Beyond reserves: options for achieving nature conservation objectives in rural landscapes. In *Frontiers in ecology: building the links.* (Eds N. Klomp, and I. Lunt.) pp. 155–168. (Elsevier Science: Oxford.)

Booth, C.A., and Burgin, S.M. (1997). Integrated catchment management linking policy, science and the community: a case study from the Upper Parramatta River Catchment, New South Wales (Australia). In *Frontiers in ecology: building the links.* (Eds N. Klomp, and I. Lunt.) pp. 29–37. (Elsevier Science: Oxford.)

Bowmer, K. (1999). Water and landscapes: perceptions and expectations. In *Preserving rural Australia: issues and solutions.* (Eds A.I. Robertson and R. Watts.) (CSIRO Publishing: Canberra.)

Cadwallader, P.L. (1978). Some causes of the decline in range and abundance of native fish in the Murray-Darling river system. *Proceedings of the Royal Society of Victoria* **90**, 211–224.

Close, A. (1990). The impact of man on the natural flow regime. In *The Murray.* (Eds N. Mackay, and D. Eastburn.) pp. 61–74. (Murray-Darling Basin Commission: Canberra.)

Commonwealth of Australia (1992). *National strategy for ecologically sustainable development.* (Australian Government Publishing Service: Canberra.)

Commonwealth of Australia (1996). *National strategy for the conservation of Australia's biological diversity.* (Australian Government Publishing Service: Canberra.)

Cullen, P., and Lake P.S. (1995). Water resources and biodiversity: past, present and future problems and solutions. In *Conserving biodiversity: threats and solutions.* (Eds R.A. Bradstock, T.D. Auld, D.A. Keith, R.T. Kingsford, D. Lunney, and D.P. Sivertsen.) pp. 115–125. (Surrey Beatty and Sons: Chipping Norton.)

Curtis, A.L. (1997). Landcare, stewardship and biodiversity conservation. In *Frontiers in ecology: building the links.* (Eds N. Klomp and I. Lunt.) pp. 143–153. (Elsevier Science: Oxford.)

Curtis, A.L., and De Lacy, T. (1995). Landcare in Australia: does it make a difference? *Journal of Environmental Management* **45**, 1:20.

Dovers, S.R., and Mobbs, C.D. (1997). An alluring prospect? Ecology, and the requirements of adaptive management. In *Frontiers in ecology: building the links.* (Eds N. Klomp, and I. Lunt.) pp. 39–52. (Elsevier Science: Oxford.)

Environment Protection Authority (1997). *Proposed interim objectives for NSW waters: inland waters.* (Environment Protection Authority: Chatswood.)

Fairweather, P.G. (1998). Agricultural chemicals and irrigation schemes what have we learnt so far? In *Free-flowing river: the ecology of the Paroo River.* (Ed. R.T. Kingsford.) (New South Wales National Parks and Wildlife Service: Sydney.)

Fairweather, P.G., and Napier G. (1998). Environmental indicators for national state of the environment reporting — Inland waters. Australia: State of the Environment (Environmental Indicator Reports). (Department of the Environment: Canberra.)

Farley, R., and Toyne, P. (1989). A national land management program. *Australian Journal of Soil and Water Conservation* **11**(2), 6–9.

Fletcher, A.R. (1986). Effects of introduced fish in Australia. In *Limnology in Australia.* (Eds D. DeDecker, and W.D. Williams.) (CSIRO Publishing: Melbourne.)

Gardner, G., and Perry, J. (1995). Dam starts up. In *Vital signs: the trends that are shaping our future.* (Eds L.R. Brown, N. Lenssen, and H. Kane.) pp. 124–125. (Earthscan: London.)

Gaston, K.J. (1996). *Biodiversity: a biology of numbers and difference.* (Blackwell Science: Cambridge.)

Gehrke, P.C., and Harris, J.H. (1994). The role of fish in cyanobacterial blooms in Australia. *Australian Journal of Marine and Freshwater Research* **45**(5), 905–915.

Gehrke, P.C., Brown, P., Schiller, C.B., Moffat, D.B., and Bruce, M. (1995). River regulation and fish communities in the Murray-Darling River system, *Australia. Regulated Rivers Research and Management* **11**, 363–375.

Harris, J.H. (1997). Environmental rehabilitation and carp control. In *Controlling carp: exploring the options for Australia.* (Eds J. Roberts, and R. Tilzey) (CSIRO Land and Water: Canberra.)

Harris, J.H., and Gehrke, P.C. (Eds) (1997). *Fish and rivers in stress. The NSW Rivers Survey.* (NSW Fisheries Office of Conservation: Cronulla, NSW.)

Healey, M., Thompson, D., and Robertson, A.I. (1997). Amphibian communities associated with billabong habitats on the Murrumbidgee floodplain, Australia. *Australian Journal of Ecology* **22**(3), 270–278.

Horwitz, P. (1994). Distribution and conservation status of the Tasmanian giant freshwater lobster *Astacopsis gouldi* (Decapoda: parastacidae). Biological Conservation **69**, 199–206.

Keenan, C., Watts, R., and Serafini, L. (1996). Population genetics of golden perch, silver perch and catfish within the Murray-Darling Basin. In *Proceedings of the 1995 Riverine Environment Forum.* (Eds R.J. Banens, and R. Lehane) pp. 17–26. October 1995, Attwood, Victoria. (Murray-Darling Basin Commission.)

Kingsford, R., and Porter, J. (Eds) (1997). *Summary of Proceedings of the Paroo River Scientific Workshop.* (New South Wales National Parks and Wildlife Service: Sydney.)

McAllister, D.E. (1991). What is biodiversity? *Canadian Biodiversity* **1**, 4–6.

MacPhail, I., and Young, E. (1992). Water for the environment in the Murray-Darling Basin. In *Water allocation for the environment.* (Ed. J.J. Pigram, and B.P. Hooper.) pp. 191–210. (Centre for Water Policy Research: Armidale, Australia.)

Mallen-Cooper, M., Stuart, I., Hides-Pearson, F., and Harris, J. (1996). Fish migration in the River Murray and assessment of the Torrumbarry fishway. In *Proceedings of the 1995 Riverine Environment Forum.* (Eds R.J. Banens, and R. Lehane.) pp. 17–26. October 1995, Attwood, Victoria. (Murray-Darling Basin Commission: Canberra.)

Miller, R.R., Williams, J.D., and Williams, J.E. (1989). *Extinctions of North American fishes during the past century. Fisheries* **14**, 22–38.

Moyle, P.B., and Leidy, R.A. (1992). Loss of biodiversity in aquatic ecosystems: evidence from fish faunas. In *Conservation biology: the theory and practice of nature conservation, preservation and management.* (Eds P.L. Fiedler, and S.K. Jain.) pp. 127–169. (Chapman and Hall: NY.)

Murray-Darling Basin Commission (1995). *Draft floodplain wetlands management strategy for the Murray-Darling Basin.* (Murray-Darling Basin Commission: Canberra.)

Murray-Darling Basin Ministerial Council (1996). *Setting the cap: report of the independent audit group.* (Murray-Darling Basin Commission: Canberra.)

New South Wales Department of Land and Water Conservation (1997). *Water reforms: securing our water future.* (NSW Department of Land and Water Conservation: Sydney.)

New South Wales Fisheries (1998). *Freshwater fishing.* http://www.fisheries.nsw.gov.au/recreational/fresh.html 24 June 1998. (NSW Fisheries.)

New South Wales National Parks and Wildlife Service (1997). *Draft NSW biodiversity strategy.* (New South Wales National Parks and Wildlife Service: Hurstville, Sydney.)

Noss, R.F. (1990). Indicators for monitoring biodiversity: a hierarchical approach. *Conservation Biology* **4**, 355–364.

Noss, R.F., and Cooperider, A.Y. (1994). *Saving nature's legacy: protecting and restoring biodiversity.* (Island Press: Washington.)

Pigram, J.J., and Hooper, B.P. (1992). *Water allocation for the environment.* Proceedings of an international seminar and workshop. (Centre for Water Policy Research: Armidale, Australia.)

Reynolds, L.F. (1976). Decline of the native fish species in the River Murray. *S.A.F.I.C.* **8**, 19–24.

Roberts, J., and Sainty, G. (1996). Listening to the Lachlan. (Sainty and Associates: Potts Point, Austalia.)

Robertson, A.I. (1997). Land-water linkages in floodplain river systems: the influence of domestic stock. In *Frontiers in ecology: building the links.* (Eds N. Klomp, and I. Lunt.) pp. 207–218. (Elsevier Science: Oxford.)

Robertson, A.I., Boon, P., Bunn, S., Ganf, G., Herzceg, A., Hillman, T., and Walker, K.F. (1996). *A scoping study into the role, importance, sources, transformations and cycling of carbon in the riverine environment.* Final Report on Natural Resource Management Strategy Project No. R6067. (Murray-Darling Basin Commission: Canberra.)

Saunders, D., Beattie, A., Eliott, S., Fox, M., Hill, B., Pressey, B., Veal, D., Venning, J., Maliel, M., and Zammit, C. (1996). Biodiversity. In *Australia: state of the environment 1996.* (CSIRO Publishing: Melbourne.)

Saunders, D.A., Hopkins, A.J.M., How, R.A. (Eds) (1990). Australian ecosystems: 200 years of utilisation, degradation and reconstruction. *Proceedings of the Ecological Society of Australia* **16**.

Saunders, D., Margules, C., and Hill, B. (1998). *Environmental indicators for national state of the environment reporting — biodiversity. Australia: State of the Environment (Environmental Indicator Reports).* (Department of the Environment: Canberra.)

Smith, P., and Smith, J. (1990). Floodplain vegetation. In *The Murray.* (Eds N. Mackay, and D. Eastburn.) pp. 215–228. (Murray-Darling Basin Commission: Canberra.)

Soule, M.E. (1991). Conservation tactics for a constant crisis. *Science* **253**, 744–749.

Stafford-Smith, M., Morton, S., and Ash, A. (1997). On the future of pastoralism in Australia's rangelands. In *Frontiers in ecology: building the links.*(Eds N. Klomp, and I. Lunt.) pp. 7–16. (Elsevier Science: Oxford.)

Walker, K.F. (1985). A review of the ecological effects of river regulation in Australia. *Hydrobiologia* **125**, 111–129.

Walker, K.F., Thoms, M.C., and Sheldon, F. (1992). Effects of weirs on the littoral environment of the River Murray, South Australia. In *River conservation and management.* (Eds P.J. Boon, P.A. Calow, and G.E. Petts.) pp. 270–293. (Wiley: Chichester.)

Walker, K.F., Sheldon, F., and Pukeridge, J.T. (1995). A perspective on dryland river ecosystems. *Regulated Rivers Research and Management* **11**, 85–104.

Wasson, B., Banens, B., Davies, P., Maher, W., Robinson, S., Volker, R., Tait, D., and Watson-Brown, S. (1996). Inland waters. In *Australia: state of the environment 1996.* (CSIRO Publishing: Melbourne.)

World Conservation Monitoring Centre (1992). *Global biodiversity: status of the earth's living resources.* (Chapman & Hall: London.)

Water and landscapes: perceptions and expectations

Kathleen H. Bowmer

Introduction

Myth, memory, art, advertising and the media shape our perceptions and expectations of the Australian landscape. This chapter discusses why, in efforts to improve, conserve or rehabilitate terrestrial and aquatic ecosystems, we inevitably confront that perception.

It seems that most scientists value the concept of naturalness, recognising the dangers of moving too far from the natural state. However, as will be illustrated, it seems that the community at large may be conditioned to the aesthetic changes brought about by land clearing, farming and river regulation. The community cannot be expected to recognise the gradual, insidious, hidden and cascading problems such as groundwater recharge, rising saline water tables, salinisation of streams, nutrient enrichment and loss of aquatic ecosystem diversity. Yet the values and expectations of the community are becoming increasingly important in developing policy in natural resource management.

State of the environment: the value of naturalness

Recent audits (Department of the Environment, Sport and Territories 1996) reported that many of our rivers and wetlands, estuaries and coastal regions are threatened by agricultural pollution and urban and industrial development. It is recognised that deterioration in water quality is often caused by changes in the hydrological cycle caused by land clearing, impoundment of water in dams and storages, and river diversion for irrigation (Bowmer 1998).

Many of the deleterious changes in Australia are inevitable consequences of agricultural developments on old and fragile soils and ecosystems adapted to dry or highly variable rainfall (Williams *et al.* 1998). Loss or fragmentation of natural habitat is a worldwide problem (Dobson 1996).

There is a concern that the changes may together reach critical thresholds, beyond which river health might be seriously degraded; and that a holistic approach to landscape and river management is required to achieve the ideals of sustainability and biodiversity which underpin the concept of a healthy river (Commonwealth Environment Protection Agency 1992).

Ecological theories which support this holistic approach include ideas of the river as a longitudinal continuum, with important lateral connections to the floodplain (Mussared 1997). The latter is important for maintaining food chains, exchange of energy and nutrients, and to allow the riverine system to act as a natural filter. Disruption to the natural hydrological cycle will disadvantage native species and communities which are well-adapted to cope with Australia's variable seasons and climates, and advantage pest plants and animals such as willows, carp and blue-green algae.

A few years ago, in the context of gathering background information for developing environmental flow in rivers of the Murray Darling Basin, I interviewed a dozen leading Australian aquatic ecologists about their views on sustainability (Young *et al.* 1995). It was an attempt to capture their perceptions of the status of riverine and flood plain ecology, and to discover how river flows might be adjusted or manipulated to move to an improved future position. However, in spite of persistent attempts to avoid retrospectivity, and instead to project the impact of alternative management options into the future, I found that most ecologists intuitively or explicitly reverted to a comparison with previous experience, using the natural or pristine state as a benchmark. Even though we recognise that return to pre-European conditions is impossible, it seems that our capacity to predict the effects of changes in natural resource management is firmly based on our experience of the past.

Recently the philosophy of moving towards a pre-existing condition or state of naturalness has been explicitly embedded in the New South Wales water reforms process. A series of river flow objectives (Environment Protection Authority 1997) is based on mimicking the rivers' natural low flows and restoring some of the natural flow variability occurring prior to regulation.

Water flow data are available for some rivers and catchments, or can be modelled and simulated, and there is some data on the origin of river sediments from channel and gully erosion derived from analysis of radionuclide tracers (Wallbrink and Murray 1993; Wallbrink *et al.* 1996). However, there are few other scientific records of the status of river or catchment health extending back beyond one or two decades. Fortunately, insights on rural landscapes can be obtained from a rich range of sources including photographs, art collections and memory.

Historical records and individual interpretation

The interpretation of the ecological status of rivers and landscapes depicted in photographs, art and memory will vary with the individual's own experience and values. For example, a magnificent oil 'Erosion Gully and White Gums', by John Wilson, hangs proudly in the Grange, the headquarters of Charles Sturt University at Bathurst. For the Vice-Chancellor, it is a valuable, precious and beautiful oil painting. However, for me it spells soil erosion, nutrient enrichment and algal blooms. It is the antithesis of sustainability.

Another item in the collection of Charles Sturt University is a black and white photograph recording the history of agricultural development on Houlaghan's Creek, a tributary of the Murrumbidgee (Figure 5.1). Ecologists today would decry the absence of riparian vegetation and note the abundance of dead trees with horror. However, a fine herd of animals was the photographer's focus then.

Figure 5.1   Houlaghan's Creek, from the Charles Sturt University Collection.

The ubiquitous presence of cattle in the Australian landscape is also evident from well-known Australian paintings. Arthur Streeton's 'Still Glides the Stream and Shall Forever Glide' (1890) and Arthur Boyd's 'Cattle on Hillside, Shoalhaven' (1975) are only two examples, spanning nearly a century. Some may regard the grazing animals as picturesque, and adding a sense of rural tranquillity. However, aquatic ecologists are now beginning to record the damaging effect of domestic stock in disturbance and in disruption of the links between a river and its floodplain (eg Robertson 1997).

Recently, Whitford (1998) gave a quantitative analysis of cattle grazing practices on the floodplain of the Murrumbidgee using data gleaned from Charles Sturt University Regional Archives, station diaries, and a range of other local sources. The history of pastoral development will be useful in assessing the extent and history of grazing impact on riverine ecology and water quality. Several stages are described: ringbarking and shrub removal between about 1860 and 1900, development of a squattocracy based on natural grasslands followed by government-sponsored closer settlement schemes until 1918, and soldier-settlement until 1939. In recent decades stocking rates increased dramatically as water became more readily available.

More recently, in a newspaper article titled 'Threatened billabongs head the way of the swagmen' (Cribb 1996), Professor Alistar Robertson of Charles Sturt University comments that 'the billabong, for nine million years the primary fount of life for the southern riverine plains, is under acute threat from carp, cattle and flood control', and also that 'the health of aquatic vegetation is a key indicator — and in many places it has completely gone'. Professor Robertson's appreciation of the degraded status of billabong ecology reflects a personal reference to previous experience and memory. He says 'when I returned to the Riverina after some years working in North Queensland, I was shocked at the extent of the degradation — not just of billabongs, but of the whole landscape'.

Other rich sources of information are the memories of people who lived, worked and grew up in rural Australia. *Listening to the Lachlan* (Roberts and Sainty 1996) is an oral history, a book of memories stretching back over the last 80 years. This is a record of particular significance because downstream the Lachlan spreads out into a large wetland, the Great Cumbung Swamp. So, unlike many other rivers, including the Darling and Murrumbidgee, the Lachlan was isolated from the Murray mainstream and was largely unaffected by paddlesteamers, bridges, large weirs, locks, desnagging and timber harvesting.

Many of the interviews suggest a degradation in water quality, and decline in native fish. Sadly, a Hillston resident remarks, 'Nowadays the river has lost its charm. It is no longer a sweet-smelling place'.

In an interpretation of the oral records Roberts and Sainty (1997) comment that over 70 years there is a clear decline in biodiversity and loss of habitat. Some changes began before carp were established in the river. Although the authors are careful not to draw any conclusions about cause and effect, they comment that loss of river water plants, particularly ribbonweed (*Vallisneria americana*), the introduction of carp, and changes in water quality are correlated.

While many natural resource managers and ecologists would regard a partial return to pristine rivers and landscapes as a worthy goal, many agriculturalists and rural industries are keen to demonstrate the advantages of economic prosperity associated with agribusiness. For example, in a recent rural press advertisement from the Ricegrowers' Association, a hot air balloon floats over a green landscape of rice paddies, irrigated cropping and associated rural industry. Under a headline 'It's a nice picture and we're going to keep it that way' and a logo proclaiming 'agricultural excellence and sustainability' the Rice Growers adopt as their benchmark the words of the explorer, John Oxley: 'the whole country seems burnt up with a long continued drought. A country which for bareness and desolation has no equal. I am the first white man to see it and I think I will undoubtedly be the last. There is a uniformity of barren desolation in this country that wearies one more than I am able to express'.

Current status: indicators and catchment health

So there are conflicting perceptions of what our rivers and landscapes should look like. However, in recent times there are opportunities to bridge the differences between the environmental movement and agricultural interests in Australia through community-led catchment governance. The concept of sustainability has been adopted by all levels of government with substantial investment in assessing the state of the environment (Department of the Environment, Sport and Territories 1996) and in searching for indicators of catchment and river health (Walker and Reuter 1996; Fairweather and Napier 1998). There is also an appreciation of the complexity of causes of degradation, and of the need to develop a holistic approach to solutions (Pratley and Robertson 1998).

For example, under a Sydney newspaper headline *The skeletons in Australia's backyard* (Woodford 1996) a researcher gazes from a scorched paddock to a bleached and leafless skeleton of a tree silhouetted against the sky. Associate Professor David Goldney of Charles Sturt University's Environmental Studies Unit estimates that in central New South Wales '... in the Molong Area alone between 15 and 28 million trees will perish from dieback caused by stress resulting from a number of factors including rising groundwater, the drift of herbicides, fungal and insect attack. And it's the same across the State'.

Commenting on the remnant of a lifeless tree he continues: 'Going to rural Australia could soon be like going to Macdonald's restaurant — you could be anywhere in the world. We are losing the landscape.' He refers to loss of diversity, and its importance in sustainability, now recognised by scientists and governments everywhere.

Spiritual and cultural values

The importance of the spiritual and cultural issues which underpin the value of place has been underestimated by policy makers until recently. The Wik land rights claim and the Yorta Yorta claim to water access in the southern Murray-Darling Basin could have far-reaching implications for water quality management, and policy on environmental flows for restoration of healthy rivers.

Aboriginal people, artists, photographers, poets and children are often particularly sensitive to the role of river and landscape; the intertwining of spiritual and physical health; and the links between past ancestors, inhabitants and future generations. Their commentary describes the landscape as a reflection of our stewardship in the past or as a resilient life support system for the future. Rivers and landscapes are seen as symbolic of life itself.

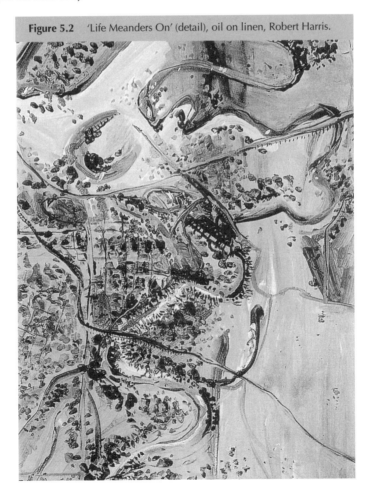

Figure 5.2   'Life Meanders On' (detail), oil on linen, Robert Harris.

Some find inspiration in landscapes as they find them today — erosion, gullies, willows, other alien species, and urban development notwithstanding. For example, Mr Robert Harris, in his recent exhibition 'The River', at Wagga Wagga City Art Gallery depicts the Murrumbidgee, its seasons, its bridges and its dreaming (Figure 5.2). 'The river can present itself as a whimsical line, dividing and scarring, or as a series of tortured intestinal turnings; as a soothing, dabbing finger benignly touching scabs of earth, or as a stern, sweeping hand re-establishing the order of the floodplain … In Wagga Wagga our sense of place requires that the watercourse, abused or benign, succulent or threatening, always be there. It is unthinkable that it should not be' (Brennan 1996).

The River Mural, produced in Wagga Wagga by schoolchildren under the tutelage of Robert Harris, is another example of the river as a special place. Harris describes the process of production and the thoughts of the children: 'the idea of a river as a divider and connector, … of flow of energy and time … change of season, floods, droughts … communities … billabongs and islands as reflections of life and relationships'.

Another Charles Sturt University artist, Treahna Hamm, is also inspired with images of river and landscape based on the importance of place. Hamm describes the concept of 'Coming into Being' as 'reflecting the two-fold significance of the unique heritage site which is in the middle of the old Wiradjuri tribal heartland and is now the middle of the old central business district of Wagga Wagga. It is a low ancient sandhill which was once part of the Murrumbidgee river system and was used as a camping and ceremonial burial site for the Wiradjuri people …'. She says, 'The intricate theme and dynamics of the print do not seek to segregate the past or cultures, but merely to marry and highlight the dual significance of the site to both black and white Australians'.

In 'Paradise Overkill', (Figure 5.3) where the overall shape of the river resembles a Koori fish trap, Hamm describes her reaction to the planned building of a tourist resort on the Murray River.

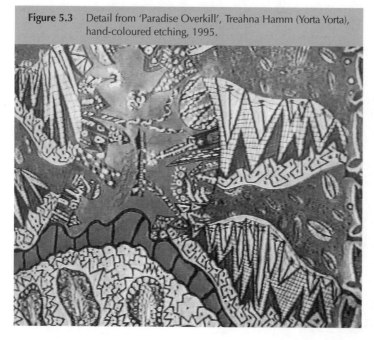

Figure 5.3   Detail from 'Paradise Overkill', Treahna Hamm (Yorta Yorta), hand-coloured etching, 1995.

'Hamm has used patches of colour to evoke the lights, excitement and bizarreness of the resort as well as the conflict between the cities along the river and the displaced bush and trees' (Johnson 1995).

So, for many, water and landscape are becoming more than an issue of natural resource management, biology and physics. In searching for the future that we would like and which is achievable, it seems that we must first revisit the past, and reconsider the landscape as a symbol of our identity.

Tim Flannery (1995) in his book *The future eaters* says 'increasingly … Australians are looking towards the bush … From a biological perspective, this is a good thing, for hopefully it will build strong links based upon understanding between Australians and their environment. Such links are vitally important, for Australians are caretakers of a disproportionately large share of the world's biological riches and Australia is a land which tolerates few mistakes. If Australians do not possess a culture which values these things they will be lost to the world …'

It is now time to plan for the future, and to agree on expectations about water and landscapes: whether to accept further deleterious changes, maintain the current status, or look for improvements.

In Australia we have moved far away from the ideal of naturalness as we have exploited our natural resources for production. Most ecologists recognise that a return to the pristine natural state is impossible, and some believe that degradation is so advanced that intervention is too late.

There may be a middle ground of rehabilitation and restoration as demonstrated by recent initiatives. For example, in 1995 the Council of Australian Governments agreed on a cap or freeze on water diversions in the Murray-Darling Basin (Murray-Darling Basin Ministerial Council 1996) and in August 1997 the government of New South Wales established a water reform package in which community-based committees were required to develop new rules on allocation of water for the environment. Other initiatives include restoration of vegetation, groundwater protection and adoption of best practice through community-based Landcare movements and land and water management plans.

Many of these initiatives build in principles of adaptive management where decisions become learning experiences for review and refinement (Grayson et al. 1994; Dovers and Mobbs 1997).

As issues become more complex and controversial, community participation and involvement of stakeholders is leading to conflict over what is acceptable, what is desirable, and who should pay, even though there is general agreement that long-term sustainability and stewardship is required.

Many community-based committees are now developing their own procedures for consultation and decision-making, and recognise the difficulty of comparing short-term quantitative economic impact data with long-term qualitative and complex environmental consequences, as well as with individual perceptions of spiritual and cultural value. The need to ensure that environmental interests are represented on important regional bodies and catchment committees was recently highlighted (Curtis 1997).

In this context, myth, memory, art, advertising and the media are powerful both in shaping our expectations and in providing a historical perspective on the rate and direction of change, but we must also be careful to look to the future, to engage with ecological concepts including biodiversity, and to guard against the hidden and insidious changes which might be difficult to reverse.

Acknowledgements

I thank the staff and students of Charles Sturt University for expanding my horizons on water and landscapes.

References

Bowmer, K.H. (in press). Water: quantity and quality. In *Agriculture and the Environmental Imperative*. (Eds J.E. Pratley, and A.R. Robertson.) pp. 36–39. (CSIRO Publishing: Melbourne.)

Brennan, M. (1996). *The river*. Paintings from the Sister Cities Project by Robert Harris, Wagga Wagga City Art Gallery, Catalogue.

Commonwealth Environment Protection Agency (1992). *Towards healthier rivers*. (Commonwealth Environment Protection Agency: Canberra.)

Cribb, J. (1996). Threatened billabongs head the way of the swagmen. *The Weekend Australian*, 27–28 April, p. 7.

Curtis, A.L. (1997). Landcare, stewardship and biodiversity conservation. In *Frontiers in ecology. Building the links*. (Eds N. Klomp, and I. Lunt.) pp. 143–153. Elsevier Science: Oxford).

Department of the Environment, Sport and Territories (1996). *State of the environment, Australia*. (Department of the Environment, Sport and Territories, CSIRO: Australia.)

Dobson, A.P. (1996). *Conservation and biodiversity*. (Scientific American Library: New York.)

Dovers, S.R., and Mobbs, C.D. (1997). An alluring prospect? Ecology, and the requirements of adaptive management. In *Frontiers in ecology. Building the links*. (Eds N. Klomp, and I. Lunt.) pp. 39–52. (Elsevier Science:.)

Environment Protection Authority (1997). *Proposed interim environmental objectives for NSW waters: inland rivers*. (Environment Protection Authority: Chatswood, NSW.)

Fairweather, P., and Napier, G. (1998). *Environmental indicators for national state of the environment reporting — waters. Australia: State of the Environment (Environmental Indicator Reports)*. (Department of the Environment: Canberra.)

Flannery, T. (1995). *The future eaters, an ecological history of the Australasian lands and people*. (Reed Books: Port Melbourne.)

Grayson, R.B., Doolan, J.M., and Blake, T. (1994). Application of AEAM (adaptive environmental assessment and management) to water quality in the La Trobe River Catchment. *Journal of Environmental Management* **41**, 245–258.

Johnson, H. (1995). *Coming into being, Treahna Hamm*. (Wagga Wagga City Art Gallery: Wagga Wagga.)

Murray-Darling Basin Ministerial Council (1996). *Setting the cap: report of an independent audit group*. (Murray-Darling Basin Ministerial Council: Canberra.)

Mussared, D. (1997). *Living on floodplains*. (Cooperative Research Centre for Freshwater Ecology and the Murray-Darling Basin Commission:.)

Pratley, J. and Robertson, A. (eds) (1998). *Agriculture and the environmental imperative*, 263 pp. (CSIRO Publishing: Melbourne.)

Roberts, J., and Sainty, G.K. (1996). *Listening to the Lachlan*. (Sainty and Associates: Potts Point, Sydney.)

Roberts, J., and Sainty, G.R. (1997). *Oral history as a tool in historical ecology: Lachlan River as a case study*. Consultancy Report 97-20, Part of NRMS Project R5058, (CSIRO Land and Water: Canberra.)

Robertson, A.I. (1997). Land-water linkages in the floodplain river systems: the influence of domestic stock. In *Frontiers in Ecology*. (Eds N. Klomp, and I. Lunt.) pp. 207–218. (Elsevier Science: Oxford.)

Walker, J., and Reuter, D.J. (1996). *Indicators of catchment health*. (CSIRO: Australia.)

Wallbrink, P.J., and Murray, A.S. (1993). The use of fallout radionuclides as indicators of erosion processes. *Journal of Hydrological Processes* **7**, 297–304.

Wallbrink, P.J., Olley, J.M., Murray, A.S., and Olive, L.J. (1996). The contribution of subsoil to sediment yield in the Murrumbidgee River basin, NSW, Australia. In *Erosion and sediment yield: global and regional perspectives*, Proceedings of the Exeter Symposium, 1996. (Eds D.E. Walling, and B.W. Webb.) pp. 347–356. (International Association of Hydrological Sciences Press: Wallingford, UK.)

Whitford, T. (1998). *An historical analysis of cattle grazing practices on the flood plain of the Murrumbidgee River*. Report of a 1997–98 Charles Sturt University Summer Vacation Research Scholarship.

Williams, J., Hook, R.A., and Gascoigne, H.L. (Eds) (1998). *Farming action catchment reaction: the effect of dryland farming on the natural environment*. (CSIRO Publishing: Australia.)

Woodford, J. (1996). The skeletons in Australia's backyard. *Sydney Morning Herald*, 6 April 1996, p. 1.

Young, W.J., Davis, J.R., Bowmer, K.H., and Fairweather, P.G. (1995). *The feasibility of a decision support system for environmental flows*. Consultancy Report No. 95/19, Final Report to the Murray-Darling Basin Commission, NRMS Project No. 5060. (CSIRO Division of Water Resources: Canberra.)

Sustaining natural resources and biological diversity in terrestrial ecosystems of rural Australia

Tony W. Norton

Introduction

The Australian continent has been occupied by humans for many tens of thousands of years. Over this period, terrestrial ecosystems and their biota have been subjected to many impacts and disturbances (Boyden *et al.* 1990). However, many of the most dramatic changes have occurred since European occupation of the continent in 1788. In a little over 200 years, most terrestrial and aquatic ecosystems have been significantly modified by human activities (Saunders *et al.* 1987; Benson 1991). Intensive land uses such as agriculture and forestry now exploit a substantial amount of the land area of rural Australia. The nation has developed extensive networks of agricultural production and is ranked amongst the world's leading producers and exporters of foodstuffs, natural fibres and livestock (Commonwealth of Australia 1992; Norton 1997). The continent presently supports over 18 million people and Australian primary production feeds and clothes in excess of 60 million people living overseas (State of the Environment 1996).

While Australia is currently recognised as one of only 12 biologically mega-diverse nations (Common and Norton 1992), its record for conserving biological diversity (the diversity of life-forms and their associations) is poor. The rate of extinction of biota over the past two centuries is high by world standards (World Conservation Monitoring Centre 1992), and at least 97 plants, 17 mammals and three species of birds have been extirpated (Boyden *et al.* 1990; Common and Norton 1992). Most of these species were associated with eucalypt-dominated ecosystems in rural Australia and have become extinct as a result of human activities. Presently, several hundred vertebrates, several thousand plants and an untold number of invertebrates are known to be or considered likely to be threatened (State of the Environment 1996; Norton 1997).

During the 1990s, Australian governments have embraced the concept of ecologically sustainable development. Stopping the loss of biological diversity, attempting to restore degraded ecosystems, and using eucalypt forests and woodlands on a sustainable basis have become higher-order social goals as indicated by the Australian government's commitment to international and domestic policies such as the 1992 United Nations Convention on Biological Diversity (UNCBD), the 1992 Australian

National Forest Policy Statement and the 1995 National Strategy for the Conservation of Australia's Biological Diversity (Commonwealth of Australia 1994). The latter policy, for example, fulfils the requirement of Article 6 of the UNCBD and was established through a national conference and other mechanisms involving major stakeholders including the National Farmers Federation and community groups from rural Australia. The national strategy outlines explicit biodiversity conservation goals to be achieved by 2000 AD and 2005 AD (Table 6.1). These goals span topics like the establishment of a national system of protected areas, the implementation of a national endangered species strategy, and the development of a program for long-term monitoring of biological diversity. While none of these goals is unique from an international perspective, they do suggest an enhanced commitment by government to tackle complex conservation and management issues that require significant, ongoing attention and funding. Even so, achieving these goals appears unlikely to be straightforward, given the current political environment of economic rationalism and the downturn in key markets for Australian industry as a result of the recent collapse of the national economies of several major trading partners in south-east Asia (Dovers and Williams 1997).

Table 6.1   Some explicit goals identified in the National Strategy for the Conservation of Australia's Biological Diversity for 2000 AD and 2005 AD (Commonwealth of Australia 1994).

2000 AD

- identification of biogeographical regions;
- identification and description of major ecosystems within each region;
- identification of a national system of protected areas representing major ecosystems;
- management plans for protected areas of major conservation significance implemented;
- the endangered species strategy implemented;
- a co-ordinated program for long-term monitoring of biodiversity implemented;
- an analysis of existing scientific knowledge conducted and gaps and priorities identified.

2005 AD

- furtherance of some of the above;
- control of three introduced mammals, 10 plants and one pathogen that pose threats to biodiversity;
- rehabilitation of 10 endangered species;
- establishment of 'voluntary or co-operative' conservation reserves.

In this chapter I discuss the conservation of biological diversity in rural Australia and consider some issues that need to be addressed if natural resources are to be used on a sustainable basis. I do not cover issues such as soil degradation and salinity which are covered elsewhere in this volume. Many of the issues that I identify suggest the need for socio-economic changes. Detailed discussion of this dimension is beyond the scope of this paper and will be covered in a later contribution. Below, I first consider the ecosystems and resources of concern and the institutional and policy frameworks in place to manage them. Then I outline some of the key issues, challenges and possible ways forward relating to the maintenance of natural resources and biodiversity in rural communities.

The settings

Forest and woodland ecosystems in rural Australia

For tens of thousands of years the Aboriginal inhabitants of Australia pursued a largely hunter-gatherer mode of existence (Dingle 1988). While information on the intensity of use of ecosystems in space and time is limited, as is detailed knowledge of regional variations in Aboriginal population size and demand for resources, a number of activities are clear. Eucalypt forests and woodlands (in what we now call rural Australia) were exploited by Aborigines for a wide range of products. Fire was an important management tool, although the extent to which burning by Aborigines affected these systems is debated (Jones 1975; Gill 1977; Williams and Gill 1995). Jones (1969) proposed that Aborigines extended the area of their preferred landscape of open-forests and woodlands by burning vegetation on a regular and systematic basis. At the time of European settlement, Aborigines deliberately lit fires for a range of reasons, including signalling, to clear the ground to assist travel and hunting, and to regenerate plant foods.

The arrival of European settlers in Australia dramatically changed the lifestyle of the Aborigines. This, in turn, affected the dynamics and potential evolution of many eucalypt forest and woodland ecosystems. The Aboriginal population was devastated by the spread of newly introduced diseases, by reductions in kangaroo, emu and possum numbers (potential food), and by direct conflicts with the Europeans. European usurpation of land proceeded rapidly between 1830 and 1900. Settlement brought rapid, fundamental and permanent changes to landscapes and the land cover (Nix 1981; AUSLIG 1990; Benson 1991; Commonwealth of Australia 1995). The vegetation of a number of regions in eastern and south-eastern Australia was permanently modified by the early 1900s. Agricultural activity in many of these areas is now extensive and the natural vegetation has largely been replaced by tussock grassland, sown pasture and crops. Sown pastures, for example, form the understorey of 50 000 km^2 of open-woodland in south-eastern Australia (Norton 1997).

The most dramatic change in terms of loss of vegetation biomass has been that from forest and woodland to grassland or pasture. The clearing of eucalypt woodland, and of tall shrubland in drier areas, gave rise to the wheatbelt in south-eastern Australia. In the period between January 1984 and November 1990 in western New South Wales, for example, clearing licences were granted for a total area of 640 000 ha, of which approximately 150 000 ha were of ecosystems dominated by eucalypts. In terms of the total area involved — 500 000 km^2 — the clearing of woodlands has resulted in one of the most significant changes in vegetation on the continent (Commonwealth of Australia 1992). Much of this woodland has been transformed into an agricultural landscape where the remnants of the former vegetation occur as isolated, uncleared patches or narrow strips along road verges. Grassy woodland ecosystems, for example, originally had an extensive geographic range on fertile soils in eastern Australia. The woodlands dominated by White Box *Eucalyptus albens* covered vast areas from northern Victoria to southern Queensland (Prober and Thiele 1993). Now, few unmodified areas remain and White Box woodlands are one of the most poorly protected ecosystems on the continent (Prober and Brown 1994). Past impacts on eucalypt forests are not dissimilar — around half of these forests have been cleared and much of the remainder have been highly modified (Norton 1996). As with many woodland ecosystems, patterns of forest clearance and modification were not random but selectively targeted the more accessible areas of higher productivity, especially in eastern and south-eastern Australia (Braithwaite *et al.* 1993; Pressey *et al.* 1996).

Currently, no single publication covers the conservation status of the biota of ecosystems in rural Australia, although some useful overviews of the status of various biota at the continental through regional scale are available (eg Thackway 1990; Thackway and Cresswell 1993). With the exception of some groups of vascular plants and vertebrates, scientific understanding of the biogeography of most biota is remarkably limited. This means that assessments of the composition, diversity and conservation status of most ecosystems are based primarily on a few vegetation descriptors such as the floristics and structure of woody plants. Lack of adequate information on the distribution and abundance of a representative range of biota precludes the characterisation and management of these systems using a more complete suite of descriptors. Unfortunately, little of the knowledge on the distribution and abundance of species held by Aboriginal people has been documented (Ross *et al.* 1994).

Sixteen vegetation formations occur in rural Australia and are defined on the basis of associations of closed-forest, open-forest, woodland, open-woodland, shrubland, open-shrubland, closed-scrub and open-scrub (Specht *et al.* 1974; Beadle 1981; Williams and Woinarski 1997). At present, many of these formations and plant communities are poorly represented in conservation reserves (Hager and Benson 1994). For example, many species of eucalypt have a restricted geographic distribution (Hopper and Coates 1990; Prober and Brown 1994; Williams and Woinarski 1997) and the conservation status of well over 100 species of eucalypts is of major concern (Kennedy 1990; Leigh and Briggs 1992). The diversity of most fauna, especially invertebrates, found in rural Australian ecosystems is less well known than that of the flora. Overviews of the broad distribution of vertebrates such as birds and reptiles have been developed by State government management agencies and others (Woinarski and Norton 1993), but have limited utility for most management purposes. Limitations in biological knowledge mean that vegetation formations, plant communities and some single species will remain the predominant surrogates for helping to conserve biodiversity and assessing progress towards the sustainable use of natural resources in rural Australia over the next few decades.

Governance, public policy and management affecting rural Australia

Australia is a federation, with the bulk of land and resource management powers residing with the eight States and Territories. The Commonwealth, nonetheless, has significant constitutional powers of intervention, especially concerning corporations, trade and foreign affairs, and can wield financial influence. In recent years the Commonwealth has adopted a greater role in co-ordinating and integrating environment and resource policy. However, few substantial statutory or institutional bases for environmental policy are in place in Australia (Dovers *et al.* 1996) and this means that many policies exist as 'commitments' that can be easily ignored or overturned by governments of the day.

The current Federal Government, the Liberal-National Coalition, has identified over A$1 billion from the part-privatisation of the national telecommunications body (Telstra) for the establishment of a Natural Heritage Trust to fund environmental programs over 5 years. A significant proportion of these monies is directed to rural Australia where a strong focus is on programs addressing land degradation. Funding is also provided (given matching finance from State and Territory governments) for goals specified in the National Strategy for the Conservation of Australia's Biological Diversity. However, significant concerns exist in relation to the current approach by government. For example, the longer term fate of much important environmental policy remains unclear since only 5 years of funding is currently available from the Natural Heritage Trust. This means that communities groups may not be able to plan and co-ordinate activities on a longer term basis because of funding uncertainty even though they are undertaking projects dealing with revegetation and land system restoration that are likely to require resources and ongoing management over decades.

A notable feature of recent Australian environmental policy is a strong reliance on community involvement. Landcare is perhaps the most prominent example of this approach, with nearly three thousand local Landcare groups established in less than a decade (Campbell 1994; Bennett *et al.* 1995). These groups have access to government funding, and many of their activities support biodiversity conservation and contribute to the sustainable use of natural resources in rural Australia. The approach is valuable for fostering a broader base of community support for sustainable environmental management, but, again, needs longer term funding and technical support to be most effective. While enormously encouraging, the actual on-ground achievements of this activity are yet to be widely evaluated, and new insights and lessons are yet to be exposed (Bennett *et al.* 1995). Indeed, this is a feature of the Natural Heritage Trust as well and it would be appropriate for far greater attention to be directed to monitoring and evaluating program delivery in these key areas. Indeed, I believe that it is vital to review the extent to which the social, economic and ecological outcomes that are intended from major national programs such as the NHT have been achieved.

During this decade, various governments have developed and promoted land and water management plans based around 'total catchment management' (TCM) as a way to achieve sustainable resource management. These initiatives have often provided significant gains in terms of previous management regimes, but have also attracted criticism because they failed to support a broader range of natural resource management goals (for a summary see Dovers 1994; Norton 1997). For example, some commentators have argued that many TCM plans are more about sustaining soils to meet agricultural production needs than ensuring that soil function and evolution is maintained as part of an overall landscape approach designed to protect all ecosystem components including biodiversity (see Norton 1997). Irrespective of the merits of these criticisms and more recent changes in the TCM approach, there is no doubt that the effectiveness of such planning could be enhanced by greater integration of policy across program areas to ensure that policy implementation proceeds on a complementary and synergistic basis.

Conserving natural resources

Many, diverse initiatives to conserve and sustainably manage the natural resources of rural Australia have been undertaken by governments, non-government organisations, individuals and partnerships over the past decade, in particular. While the degree of success of activities has varied (and a number have failed) they have undoubtedly enhanced awareness in Australian society of the need for sustainable management, and the roles that local and regional communities can foster in the promotion of environmental stewardship and an improved land-use ethic.

Umbrella and flagship species

Use of 'umbrella species' and 'flagship species' to help conserve biological diversity at the landscape or land system level is an approach that has become popular in government policy in many countries. The rationale of the umbrella species concept is that, in the absence of comprehensive biological data, by protecting the minimum area needed for a viable population of a single, large-bodied species such as a top-order carnivore like the Australian Powerful Owl *Ninox strenua* (Norton and May 1994), sufficient space may be maintained to ensure the viability of smaller and more numerically abundant (but poorly understood) species in the same area. The concept has been considered for species such as the Grizzly Bear *Ursus arctos*, Tiger *Panthera tigris* and African Elephant *Loxodonta africana* (Michelmore 1994) and, with further development, has useful applications in rural Australia. A 'flagship species' is one that attracts sufficient public interest that it can be employed to promote community awareness and the need for conservation and the sustainable use of natural resources. Ideally, such as species would be found in the region where the conservation and management outcomes are to be delivered, and would be of interest to gain the support of most sections of the community including business and commerce.

In Australia, the Koala *Phascolarctos cinereus* has become an important flagship for promoting the conservation of biological diversity and sustainable resource use in many rural areas. Promotion of the conservation needs of the animal by groups such as the Australian Koala Foundation and the Australian Conservation Foundation has had a significant impact on diverse sections of the community including governments, business and non–government organisations. During the 1990s the species has been used to illustrate the plight of species suffering from habitat loss and non–sustainable resource developments (eg Allen 1997). A number of resource developments potentially threatening the habitat of the species have been stopped or significantly modified because of the major political profile of this species. Legislation relating to endangered species and threatening processes has been strengthened as a result of political concern for the animal (Dovers *et al.* 1996). Many community-based programs have been undertaken to plant eucalypt species that can be eaten by the Koala, and to protect remnant vegetation to provide continuity of habitat and shelter for populations (Bennett *et al.* 1995). Survey teams comprising groups of private landowners and rural residents have been established to undertake systematic and sophisticated surveys and monitoring of the species in some areas of its range (eg central Queensland, south-east NSW) where it appears threatened with extinction (Allen 1997).

Put in a broader context, changes in societal behaviour as a result of concern for the conservation of species such as the Koala have resulted in considerable reforms including the creation of proactive groups, enhanced financial support for political lobbying and environmental education, modifications in environmental legislation, revegetation and ecosystem restoration programs, and the creation of community-based biodiversity 'stewardship' groups (Campbell 1994). There is merit in identifying more flagship species and umbrella species in rural areas to support moves towards sustainable resource management in rural areas.

Some issues and challenges for rural Australia

While the reforms and the potential for further changes considered above are highly significant, the challenges confronting rural Australia should not be underestimated and include:

* eliminating processes that threaten ecosystems;
* developing a more integrated approach to conservation and land management;
* creating a national system of protected areas; and
* developing more effective government policy and institutional capacity for learning and adaptation.

Major processes that threaten terrestrial biodiversity such as the clearance of native vegetation, and broad-scale destruction and degradation of habitat have been identified for many eucalypt forests and woodlands and have been known for almost a century. Unfortunately, these processes continue in most regions of eastern and south-east Australia. For example, the present rate of clearance of native vegetation is the highest of any developed nation (Norton 1997) and contrasts with stated government policy to 'protect the bush' and restore degraded ecosystems. These practices lead to the bizarre situation where many millions of dollars are currently spent on planting trees while, at the same time and often in the same region, extensive clearance of native vegetation continues.

Extinction rates and the decline in the distribution and abundance of native biota have been greatest in the agricultural and pastoral areas, and in areas settled first by Europeans. In rural Australia the potential detrimental impact of introduced herbivores is well documented (Norton 1997). Perhaps less obvious are exotic predators such as the Fox *Vulpes vulpes* and feral Cat *Felis catus* which exploit a large range of eucalypt-dominated ecosystems, largely in the absence of effective controls. The indiscriminate use of fire, spread of disease, loss of hollow-bearing trees, and strict control of water flows and flooding regimes in waterways and rivers are further examples of the many landscape-scale processes that continue to threaten rural biota (Kirkpatrick 1994). While governments have developed and implemented many policies concerned with the protection of threatened taxa, progress on addressing threatening processes has been slow (Dovers and Williams 1997).

Australian governments have recently recognised the need for a more coherent, co-operative and integrated approach to conservation and management. The Intergovernmental Agreement on the Environment (Commonwealth of Australia 1992) defined a number of roles for governments, including promoting consistency in relevant data collection and its integration; assistance in the development of co-ordinated standards for the management of conservation reserves and the mechanisms used to select new reserves; and co-operation in conservation actions affecting rare species and conservation areas that cross State borders. Previously, I have argued that these roles could best be pursued and refined by governments through co-operative regional initiatives aimed at enhancing biodiversity assessment, and conservation planning and management (see Norton 1997). Initiatives requiring co-operation among multiple agencies and landowners exemplify the sorts of activities at the regional scale that are vital in empowering local groups and promoting the sustainable use of natural resources (Bridgewater 1994).

The need to better understand the structure and function of ecosystems to meet conservation and management objectives is widely acknowledged, and this is true of virtually all terrestrial and aquatic systems in rural Australia. The present, strong reliance on vegetation descriptors to characterise ecosystems has limitations, particularly when the target for conservation and management is not vegetation *per se*. Concepts such as indicator and keystone species and functional types have been proposed to facilitate the characterisation and tracking of the state of ecosystems, but require further evaluation if they are to be used reliably (Trueman and Cranston 1994). Agreement is required on the precise units to be used for the evaluation and management of biological diversity at the regional scale (eg Kirkpatrick and Brown 1994), and the development of mechanisms to facilitate the acquisition and management of new areas selected for reservation is required if the establishment of an adequate national system of protected areas is to be achieved (Margules 1989; Woinarski and Norton 1993). A number of community-based groups and activities have been initiated to survey and monitor species populations (eg Birds Australia, Frogwatch). Given appropriate financing and support, the expertise and skills held by these groups could be more widely employed to assess other natural resources at the regional level, especially on private lands where previous surveys are limited or absent (Common and Norton 1994).

Despite being identified as a key target in the National Strategy for the Conservation of Australia's Biological Diversity (Table 6.1), government progress on developing a broadly-based ecological monitoring program has been slow. A serious generic problem of environmental policy in rural Australia is that few ecosystem and biodiversity management prescriptions are monitored (Walker 1994). Given the imprecision of current knowledge, most such prescriptions are best treated as working hypotheses to be refined or changed as new information becomes available (Murphy and Noon 1991; McCarthy and Burgman 1995). However, adopting such an approach requires explicit operational goals for management that can be routinely evaluated.

Many areas in eastern and south-east Australia exposed to key threatening processes are in need of ecological monitoring. Systematic and on-going approaches for the mitigation of, elimination of and education about key threatening processes could be strengthened readily using the principles for sustainable resource management as an exemplar in regional communities.

Conclusions

The impact of humans on natural resources and biological diversity in rural Australia has been dramatic. Statistics inadequately convey the extent of such impacts; nor do they convey the extraordinary efforts that appear necessary to restore degraded systems and to use the remaining ecosystems on a sustainable basis. A flexible, long-term approach to conservation and management appears essential if key elements of rural Australian ecosystems are to be conserved. The approach must be landscape-based and management decisions must be complementary, irrespective of land tenure. Increased protective measures are required at all scales — regional, landscape, catchment and sub-catchment. Integrated landscape management requires that approaches to the management of reserve and off-reserve areas be conceived jointly, carefully co-ordinated and implemented in a precautionary manner.

If sustainable management is to be achieved, then a number of changes to the legislative base, design and implementation of land use prescriptions and codes of practice appear unavoidable. These changes will also have important socio-economic implications for rural Australia. Indeed, translating scientific and other understandings into effective ecosystem management will require many changes at all levels of society. While landscape- and ecosystem-scale approaches to adaptive management are poorly developed in rural Australia, the technical capacity for rapid reform does exist, given broad support and the constructive and ongoing engagement of local communities and indigenous peoples.

References

Allen, C. (1997). *Biodiversity conservation, Koala research and local communities*. Johnstone Centre Report No. 102. (Charles Sturt University: Albury, Australia.)

AUSLIG (1990). *Atlas of Australian resources*. Third series, vol. **6**, Vegetation. (AGPS: Canberra.)

Beadle, N.C. (1981). *The vegetation of Australia*. (Cambridge University Press: Cambridge, UK.)

Bennett, A., Backhouse, G., and Clark, T.W. (Eds) (1995). People and nature conservation — perspectives on private land use and endangered species recovery. *Transactions of the Royal Zoological Society of New South Wales, Sydney*.

Benson, J. (1991). The effect of 200 years of European settlement on the vegetation and flora of New South Wales. *Cunninghamia* **2**, 343–370.

Boyden, S., Dovers, S., and Shirlow, M. (1990). *Our biosphere under threat: ecological realities and Australia's opportunities*. (Oxford University Press: Melbourne.)

Braithwaite, L.W., Belbin, L., and Austin, M.P. (1993). Land use allocation and biological conservation in the Batemans Bay forests of New South Wales. *Australian Forestry* **56**, 4–21.

Bridgewater, P.B. (1994). Conservation strategy and research in Australia — how to get to the 21st century in good shape. In *Conservation biology in Australia and Oceania*. (Eds C. Moritz, and J. Kikkawa.) pp. 17–25. (Surrey Beatty and Sons, Sydney.)

Campbell, A. (1994). *Landcare — communities shaping the land and the future*. (Allen & Unwin: Sydney.)

Common, M.S., and Norton, T.W. (1992). Biodiversity: its conservation in Australia. *Ambio* **21**, 258–265.

Common, M.S., and Norton, T.W. (1994). Biodiversity, natural resource accounting and ecological monitoring. *Environmental and Resource Economics* **4**, 29–53.

Commonwealth of Australia (1992). *Australia's environment: issues and facts*. (Australian Bureau of Statistics: Canberra.)

Commonwealth of Australia (1994). *Draft national strategy for the conservation of Australia's biological diversity*. (AGPS: Canberra.)

Commonwealth of Australia (1995). *Native vegetation clearance, habitat loss and biodiversity decline*. Biodiversity Series, Paper No. 6. (Biodiversity Unit, Department of the Environment, Sport and Territories: Canberra.)

Dingle, T. (1988). *Aboriginal economy — patterns of experience*. (McPhee Gribble in association with Penguin: Melbourne.)

Dovers, S.R. (1994). *Australian environmental history: essays and cases*. (Oxford University Press: Melbourne.)

Dovers, S.R., Norton, T.W. and Handmer, J.W. (1996). Uncertainty, ecology, sustainability and policy. *Biodiversity Conservation* **5**, 1143,1167.

Dovers, S.R., and Williams, J.E. (1997). Implementing the convention on biological diversity: the Australian experience. *Ambio* (in press).

Gill, A.M. (1977). Management of fire-prone vegetation for plant species conservation in Australia. *Search* **8**, 20–26.

Hager, T.C., and Benson, J.S. (1994). *Review of the conservation status of forest vegetation communities in New South Wales. Part 3 Assessment of the conservation status of forest plant communities in north eastern NSW*. Final report to the Australian Heritage Commission, Canberra.

Hopper, S.D., and Coates, D. (1990). Conservation of genetic resources in Australia's flora and fauna. *Proceedings of the Ecological Society of Australia* **16**, 567–577.

Jones, R. (1969). Fire-stick farming. *Australian Natural History* **9**, 224–228.

Jones, R. (1975). The Neolithic, Palaeolithic and the hunting gardeners: man and land in the Antipodes. In *Quaternary studies.* (Eds R.P. Suggate and M.M. Cresswell.) (The Royal Society of New Zealand: Wellington.)

Kennedy, M. (1990). *A complete reference guide to Australia's endangered species.* (Schuster: Sydney.)

Kirkpatrick, J.B. (1994). *A continent transformed. Human impact on the natural vegetation of Australia.* (Oxford University Press: Melbourne.)

Kirkpatrick, J.B., and Brown, M.J. (1994). A comparison of direct and environmental domain approaches to planning reservation of forest higher plant communities and species in Tasmania. *Conservation Biology* **8**, 217–224.

Leigh, J.H., and Briggs, J.D. (1992). *Threatened Australian plants: overview and case studies.* (Australian Nature Conservation Agency: Canberra.)

Margules, C.R. (1989). An introduction to some developments in conservation evaluation. *Biological Conservation* **50**, 1–11.

McCarthy, M.A., and Burgman, M. (1995). Coping with uncertainty in forest wildlife planning. *Forest Ecology and Management* **74**, 23–36.

Michelmore, F. (1994). Keeping elephants on the map: case studies of the application of GIS for conservation. In Mapping the diversity of nature. (Ed. R. I. Miller.) pp. 107–124. (Chapman and Hall: London.)

Murphy, D.D., and Noon, B.R. (1991). Coping with uncertainty in wildlife management. *Journal of Wildlife Management* **55**, 773–782.

Nix, H.A. (1981). The environment of Terra Australis. In *Ecological biogeography of Australia.* (Ed A. Keast.) pp. 105–131. (Dr W. Junk, The Hague.)

Norton, T.W. (1996). Conservation of biological diversity in Australia's temperate eucalypt forests. *Forest Ecology and Management* **85**, 21–34.

Norton, T.W. (1997). Conservation and management. In *Eucalypt ecology. Individuals to ecosystems.* (Eds J.E. Williams, and J.C.Z. Woinarski.) Chapter 15. (Cambridge University Press: Cambridge.)

Norton, T.W., and May, S.A. (1994). Towards sustainable forestry in Australian temperate eucalypt forests: ecological impacts and priorities for conservation, research and management. In *Ecology and sustainability of southern temperate ecosystems.* (Eds T.W. Norton and S.R. Dovers.) pp. 10–30. (CSIRO: Melbourne.)

Pressey, R.L., Ferrier, S., Hager, T.C., Woods, C.A., Tully, S.L. and Weinman, K.M. (1996). How well protected are the forests of north-eastern New South Wales? Analyses of forest environments in relation to formal protection measures, land tenure, and vulnerability to clearing. *Forest Ecology and Management* **85**, 311–334.

Prober, S.M., and Thiele, K.R. (1993). The ecology and genetics of remnant grassy White Box woodlands in relation to their conservation. *Victorian Naturalist* **110**, 30–36.

Prober, S. M., and Brown, A. D. H. (1994). Conservation of the grassy White Box woodlands: population genetics and fragmentation of *Eucalyptus albens. Conservation Biology* **8**, 1003–1013.

Ross, H., Young, E., and Liddle, L. (1994). Mabo: an inspiration for Australian land management. *Australian Journal of Environmental Management* **1**, 24–41.

Saunders, D. A., Arnold, G. W., Burbidge, A. A., and Hopkins, A. J. M. (Eds) (1987). *Nature conservation. The role of remnants of native vegetation.* (Surrey Beatty & Sons: Sydney.)

State of the Environment (1996). *State of the environment of Australia.* (AGPS: Canberra.)

Specht, R.L., Roe, E.H., and Boughton, V.H. (1974). Conservation of major plant communities in Australia and Papua New Guinea. *Australian Journal of Botany Supplementary Series* **7**, 1–667.

Thackway, R. (1990). *Conservation of forested and wooded vegetation types in nature conservation reserves in Australia.* (Australian National Parks and Wildlife Service: Canberra.)

Thackway, R., and Cresswell, I.D. (1993). *Environmental regionalisations of Australia: a user-oriented approach.* (Environmental Resource Information Network, DEST: Canberra.)

Trueman, J.W.H., and Cranston, P.S. (1994). *An evaluation of rapid biodiversity assessment for estimating arthropod diversity.* A report to the Commonwealth Department of the Environment, Sport and Territories, Canberra.

Walker, K.J. (1994). *The political economy of environmental policy — an Australian introduction.* (University of New South Wales Press: Sydney.)

Williams, J.E., and Gill, A.M. (1995). *Impacts of fire regimes on native forests in eastern New South Wales.* New South Wales National Parks and Wildlife Service, Issues Paper No. 2. (NSW National Parks and Wildlife Service: Sydney.)

Williams, J.E., and Woinarski, J.C.Z. (1997). *Eucalypt ecology. Individuals to ecosystems.* (Cambridge University Press: Cambridge.)

Woinarski, J.C.Z., and Norton, T.W. (1993). *Towards a national system of forest reserves.* Discussion paper prepared for the Forest Unit, Commonwealth Department of the Environment, Sport and Territories, Canberra.

World Conservation Monitoring Centre (1992). *Global biodiversity — status of the Earth's living resources.* (Chapman Hall: London.)

Nutrients and algal blooms: lessons from inland catchments

Dhia Al Bakri and Mosharef Chowdhury

Introduction

This chapter presents an alternative view regarding the sources of the nutrients, particularly phosphorus (P), entering Australian inland waterways. This view has critical implications for the development of sustainable management systems to control eutrophication (nutrient enrichment) and algal blooms in rural Australia. The view is based on results from a case study carried out in the catchment of Orange water supply and on evidence from other recent studies undertaken elsewhere in Australia.

Eutrophication and the related excess algal growth is one of the major water quality problems confronting inland waterways, water supplies of country towns and rural communities in Australia (Causely, 1989; Oliver *et al.*, 1993; Buchan and Diacono, 1995). The blue–green algal blooms (cyanobacteria) have a wide range of social, economic and environmental impacts. Some species produce toxins which are harmful to humans and livestock. Other impacts include loss of recreation amenity, loss of biodiversity, and adverse impacts on aquatic biota and inland fisheries. Excess growth of blue–green algae in water supplies is aesthetically objectionable, potentially toxic and responsible for the introduction of unpleasant odour and taste to drinking water. Other problems include filter clogging and major cost implications for water treatment (Cullen 1986; Verhoeven 1993; Al Bakri *et al.* 1995). The cost of algal bloom in Australia runs into millions of dollars annually. For example, the 1991–92 losses in tourism and recreational benefits for New South Wales alone were estimated at $9.4 million (Blue-Green Algae Task Force 1992).

Factors and processes controlling algal blooms

Blue-green algal outbreaks depend upon the complex and dynamic interactions of a host of biophysical processes and socio-economic factors. The most important causal factors are:

- high nutrient levels, particularly P and nitrogen (N);

- low N:P ratios (< 29:1);

- high water temperature (> 18–20° C);

- high pH (8–10) and low carbon dioxide concentration;

- high grazing pressure from zooplanktons on phytoplanktons other than blue-green algae (blue-greens are relatively inedible);

- low flows, leading to long retention times and calm water condition;

- thermal stratification in dams and reservoirs which reflects little or no mixing between the surface water layer and bottom water layer;

- reduction in turbidity to moderate levels, leading to increased light intensity and photosynthesis by algae;

- oxygen depletion (anoxia) in the bottom water layer which is usually associated with thermal stratification. The development of anoxic conditions facilitates the release of bioavailable nutrients from the bottom sediments to the water column. (Blue-Green Algae Task Force 1992; Harris 1994).

Algal blooms have always been with us because they are an integral part of natural processes characterising degraded aquatic ecosystems. The first algal bloom was recorded as early as 1830 by the explorer Charles Sturt when he noted that the waters of Barwon Darling were showing signs of decay and a slight tinge of green. In fact, the world's first scientific record of an algal bloom was published in the journal *Nature* in 1878, and described a massive bloom in Lake Alexandria, South Africa (Carson 1992). Algal blooms, however, have been exacerbated as a result of mismanagement and degradation of our land and water resources. For instance, introduction of modern management systems and technologies to regulate our river system resulted in reducing natural flow, enhancing the potential for the development of thermal stratification and anoxia and causing imbalance in the chemical and ecological dynamics of the natural aquatic ecosystems. Intensive use and management of land have exacerbated erosion and subsequent transportation of nutrients and other contaminants to the waterways. All of the above have contributed significantly to the worsening of eutrophication and algal blooms in inland water resources.

Of the above factors, it is widely accepted that nutrient level (particularly N and P) and flow are the most important factors affecting formation of nuisance algal blooms. It is also believed that these two factors are relatively easy to control and thus provide critical tools to manage algal growth (Grace *et al.* 1997; Shafron 1995; Harris 1994). As the flow in many inland catchments, particularly during the dry period, is difficult to manipulate, the reduction of nutrient loading becomes the most crucial management tool to reduce the severity of nuisance algal blooms. It is widely accepted that P is the dominant nutrient limiting algal growth in freshwater systems whereas N is the dominant nutrient in coastal and estuarine systems. Consequently, P is considered the most critical nutrient to control algal growth in inland waterways (Hecky and Kilham 1988; Cullen 1986; Harris 1994).

The conventional view suggests that most P in Australian inland waterways comes from fertilised topsoil, removed by surface soil erosion or as agricultural runoff, and/or from point sources such as sewage and industrial effluents (Hart 1996; Harris 1995; Beck and Long 1994; Verhoeven 1993; O'Connor 1992). Based on this understanding, current strategies to manage algal blooms in rural Australia have focused on reducing P entering waterways from these sources (Verhoeven 1993; Hart

1996). The current understanding of the sources, transport and in-stream process of P is based primarily on studies carried out in the more heavily urbanised or industrialised Europe and North America (Harris 1995; Donnelly *et al.* 1996; Wasson *et al.* 1996, Grace *et al.* 1997). There is now, however, mounting evidence to suggest that the concepts and assumptions underpinning this conventional understanding are not entirely valid for many turbid Australian rivers. A number of recent studies have argued that background P in soils (native P), derived from rock weathering, represents a major source of P in many inland waterways (Murray 1996). All rocks contain small amounts of phosphate (Table 7.1), present in an insoluble form as the accessory mineral apatite, and weathering of bedrock is the ultimate source of all phosphorus in the environment (Cook 1983). Therefore, soils derived from rocks relatively rich in P tend to contain significant amounts of background P. According to Wasson *et al.* (1996) and Murray (1996), gully and stream bank erosion (subsoil sources) is believed to be more important than surface erosion in supplying sediments to many waterways in rural Australia. As the native P must be transported into waterways with sediments, the authors concluded that gullies and stream banks are major sources of particulate P. Internal loading processes were considered responsible for releasing part of the particulate P to the water column as dissolved P (Kilham and Kilham 1990; Murray 1996). The internal loading mechanism is the process of nutrient release from anoxic, and to lesser extent from oxic, bottom sediments of water bodies to the water column. As a result of several reduction/ oxidation cycles which are facilitated by sulfate-reducing bacteria, a significant portion of the particle-bound P in the bottom sediment can be reduced and released as solution in pore water (Murray 1996; Grace *et al.* 1997). The supply of dissolved P from the pore water to the water above the sediment column is controlled by molecular diffusion and the convective transfer mechanism processes. Bioturbation, water turbulence and macrophytes uptake are also some factors which influence the convective process of nutrient transfer (Gonsiorczyk *et al.*1997).

Table 7.1   Phosphate content of common rock types.

Igneous % P_2O_5		Metamorphic % P_2O_5		Sedimentary % P_2O_5	
Granitic	0.16	Gneiss	0.2	Sandstone	0.08
Intermediate	0.42	Schist	0.2	Shale	0.17
Basic	0.27	Phyllite	0.2	Limestone	0.04
Ultrabasic	0.30	Amphibolite	0.3	Clay	0.30

Source: McKelvey 1973, given in Cook 1983, p. 8.

A case study in a rural catchment
— Orange water supply catchment

Blue-green algal outbreaks have been frequently reported in the main reservoir of the Orange water supply (Suma Park Reservoir), particularly during summer and autumn.

The water supply is part of the upper Macquarie catchment and has an area of 184 km². Suma Park Reservoir is located on the upper reaches of Summer Hill Creek and has a storage capacity of 18 000 ML. The study area is characterised by a temperate climate with a mean annual rainfall of 809 mm (Kovac and Lawrie 1990). Most of the catchment is cleared and, according to Taylor (1994), the main land use categories include native or improved pasture (77%), cropping (11.9%), and horticulture (6.5%). Tertiary basalt and Ordovician igneous rocks, which occupy more than 80% of the area, are rich in the phosphate mineral apatite $[Ca_5(F,Cl)(PO_4)_3]$ (Scott *et al.* 1994; Wyborn 1994). As a result, the catchment's soils are expected to be phosphate-rich too. The catchment is very stable with a permanent vegetation cover. More than 85% of the area indicates no appreciable surface erosion and only about 15% of the catchment shows evidence of minor to moderate sheet erosion. However, substantial stream bank and gully erosion was evident. Approximately 40% of the total channel length in the catchment was showing severe to very severe bank erosion (Taylor, 1994).

A total of 300 water samples were collected from 12 sampling sites representing all the creeks and reservoirs in the catchment. The samples and field measurements were taken on a weekly basis between 3 August 1995 and 31 January 1996. During storm events additional water samples and measurements were taken. Profiling of Suma Park Reservoir was undertaken at approximately fortnightly intervals during January–May 1996 and again during January–April 1997. The presence of phosphorous and nitrogen compounds and concentration of suspended sediment were determined following the methods described in HACH (1992). The field methods to measure the physico-chemical parameters and the analytical techniques to determine nutrient concentrations and suspended sediment were described in Al Bakri and Chowdhury (1997) and Chowdhury and Al Bakri (1998).

Algae

During 1995 and 1996, levels of up to 70 000 cells/mL of cyanobacteria (*Anabaena* and *Microcystis*) were recorded in Suma Park Reservoir. These levels are significantly higher than the cyanobacteria guidelines recommended for raw waters for drinking purposes in Australia, which are 1 000–2 000 cells/mL (Australian and New Zealand Environment and Conservation Council 1992).

Phosphorus compounds

The spatial and temporal variation of the P compounds are discussed in detail in Al Bakri and Chowdhury (1997) and Chowdhury and Al Bakri (1998). The total phosphorus (TP) concentrations recorded in this study are among the highest measured in New South Wales; many of them exceed the maximum limit of the Australian guidelines (Australia and New Zealand Environment and Conservation Council 1992). Virtually all sampling sites have median TP concentrations equal to or greater than 0.05 mg/L with mean values ranging between 0.05–0.11 mg/L. During low flow, the catchment had a relatively clear water, but TP concentrations were still relatively high with median values ranging between 0.05 mg/L and 0.06 mg/L. These levels are still above the recommended value to prevent excess algal growth. During this period, 45% to 64% of TP was present as dissolved reactive P (DRP) available for algal growth. When it is considered that as little as 0.01 mg/L of DRP is sufficient for excess algal growth (CSIRO 1994), it is not surprising that excess algal growth often occurs during low flow in the catchment, when high temperature, high light availability and degraded ecosystems are dominant.

The annual TP load entering the Suma Park Reservoir ranged from a minimum of 1102 kg, according to the load duration method, to a maximum of 2843 kg, according to the regression method. Annual export coefficient was estimated to be between 0.06 kg/ha/yr and 0.16 kg/ha/yr. In comparison with estimates given in Tables I & II in Cullen and O'Loughlin (1982), P export from this catchment is considered high relative to P export from agricultural/grazing land in Australia.

Regression analysis showed that TP load data have significant positive relationships with stream discharges and suspended solid loads. The highest P loads occurred during flood when sediment loads were also high. Approximately 70% of P load and 73% of suspended solid load entering Suma Park Reservoir were transported by 70% of the flow that occurred during 10% of the study time (Al Bakri and Chowdhury 1997). This indicates that sediment-bound P (particulate P) transported to the waterways during high flow events dominates the catchment loading. The flow and sediment dependency of P load and the significant increase in P load during storm flow have been widely reported (Robinson and Hatfield 1992; Cosser 1989).

Total phosphorus concentrations in Suma Park Reservoir were found significantly higher at the bottom layer than at the surface water. The range of TP concentrations at the deepest point of reservoir (30 m) varied between 0.07 and 0.49 mg/L, whereas the TP concentrations in the surface water ranged between 0.06 and 0.10 mg/L. During thermal stratification of the reservoir, when the dissolved oxygen level was depleted, DRP concentrations in the bottom water were significantly higher (0.07–0.11 mg/L) than those at the surface layer (0.01–0.04 mg/L). This trend of DRP variation clearly indicates that the depletion of oxygen (anoxic condition) at the bottom of the reservoir enhances the release of soluble P from the bottom sediments.

Sources of phosphorus

With the exception of a minor effluent from the village Spring Hill's wastewater treatment plant, which is used for irrigation, and one dairy farm, the Orange water supply catchment is devoid of any significant point sources. The study results do not show any systematic variation in P concentration/ load in the catchment to indicate that leaks from septic tanks or any other point sources form important contributors to the P input. For example, there was no significant change in TP concentration upstream or downstream of Lucknow, which is the main urban centre in the catchment (Al Bakri and Chowdhury 1997). This lack of significant variation in P concentration with stream length confirms that input from point sources was not important in terms of total catchment load.

Evidence discussed in Al Bakri and Chowdhury (1997) and Chowdhury and Al Bakri (1998) indicates that the application rate of phosphate fertilisers in the catchment is considered low. Furthermore, most of the soil in the catchment is heavy and acidic and thus has a great potential to fix (immobilise) P fertilisers. Taking into consideration this evidence and the fact that surface soil erosion, which is the necessary means for removing P from the catchment to the waterways, is insignificant in the catchment, it can be argued that P fertiliser carried by agricultural runoff is not a primary source of P in the waterways of the catchment.

The results of this study along with evidence from other studies (Oliver et al. 1993; Caitcheon et al. 1995; Murray 1996; Donnelly et al. 1996), indicate that the background P in the naturally P-rich soils derived from Tertiary basalt and other volcanics in the catchment was the primary source of P in the catchment waterways. The P which is bound up with fine suspended sediment is transported to the waterways from gully and stream bank erosion. An internal loading mechanism is believed to play an important role in releasing bioavailable P from the particle-bound P in the bottom and suspended sediments causing eutrophication in the Suma Park Reservoir.

Nitrogen compounds

With the exception of the control site (Site 12) all sites studied have median total nitrogen (TN) concentrations greater than the upper limits of the Australian TN guidelines for preventing excess algal growth (Australia and New Zealand Environment and Conservation Council 1992). An annual TN load of at least 9946 kg, and probably as high as 27 857 kg, entered Suma Park Reservoir during February 95–January 96. TN loading was flow dependent and dominated by storm events. The concentrations of nitrate nitrogen and ammonium nitrogen formed approximately 30% and 5% of the TN concentration, respectively. Taking into account that the presence of ammoniacal nitrogen is considered to be an indicator of sewage pollution in a stream, it can be suggested that the contribution of sewage sources in the catchment is very limited. Biological fixation, atmospheric deposition, fertilisers, manure and decomposition of organic matter are believed to be the main sources of nitrogen to the catchment waterways. Point sources such as septic tanks, sewage effluents, and discharges from feedlots or dairies are not believed to be significant contributors of nitrogen in the catchment.

An important factor in the success of some blue–green algae is their ability to fix nitrogen from the atmosphere. Members of the heterocystous blue–green algae such as *Anabaena* are very efficient in N fixation. Non-heterocystous blue–green algae such as *Microcystis* also fix nitrogen but less efficiently than the former group. The nitrogen fixation capabilities of blue–green algae *Anabaena* and *Microcystis*, which are identified in the Suma Park Reservoir, enable them to overcome nitrogen deficiency in the waters (Blue–Green Algal Task Force 1992). Due to this feature and because the N cycle is very complex and difficult to control, it can be argued that the widely accepted view — that controlling P loading in the freshwater systems is the most efficient way to control algal growth — is applicable to Orange catchment (Hecky and Kilham 1988; Harris 1994).

Conclusions and management implications

Implications for Orange catchment

The study revealed that P is the most important factor limiting the algal growth in the Orange catchment. Consequently, the management of the algal problem should focus on reducing P loading (both external and internal loadings) in the water supply. Contrary to the conventional view, the anthropogenic sources (agricultural runoff and point sources) were found to be insignificant contributors, in comparison to native (background) sources, to the P input in the waterways. Particle-bound P from naturally phosphate-rich basalt and volcanic soils is transported to the streams and reservoirs from gully and stream bank erosion. The internal loading mechanism is believed to play an important role in releasing bioavailable P from the bottom and suspended sediments to the water column. Based on this understanding, catchment-wide management strategies and reduction of superphosphate fertiliser application, although useful for many conservation purposes, may not be effective in terms of reducing eutrophication and controlling algal blooms. The management of algal blooms should, therefore, focus on manipulating P and sediment within the channels and waterbodies of the catchment. The following actions and management strategies should be targeted.

Short-term actions

The short-term actions should aim to have a rapid effect by treating or controlling the effect of the algal bloom problem. This aim can be achieved by manipulating the waterbody within Suma Park Reservoir itself through effective destratification, using multi-level offtakes, protecting the water offtakes from excessive algal growth and removing toxins causing taste and odour complaints using activated carbon treatment.

Long-term strategies

The long-term strategies should aim to eliminate or reduce the causes of algal blooms by adopting planning and management activities to stop or reduce sediment and associated nutrients washing into creeks and dams. Important elements of these strategies may involve:

- controlling stream bank and gully erosion in the catchment to reduce the P and sediment getting into the stream flow;

- establishing buffer strips at critical locations on the waterways by increasing riparian vegetation to increase bank stability;

- dredging the bottom sediment in the inlet of Suma Park Reservoir to remove nutrient-rich sediments and to reduce the effect of internal loading on bioavailable P in the water column;

- establishing a CatchmentCare group to co-ordinate the efforts of the community and government agencies in developing and implementing sustainable management strategies.

Implications for rural Australia

Recent studies carried out elsewhere have also shown that native P, derived from gully and stream bank erosion of subsoils, contributes significant particulate P to many inland waterways. Caitcheon *et al.* (1995) demonstrated that the dominant source of P and fine sediments in Chaffey Reservoir comes from unfertilised naturally P-rich Tertiary basalt soils. These researchers argued that there is a strong link between intense bacterial sulfate reduction and enhanced bioavailable P release from the reservoir's bottom sediments. A study of the Murrumbidgee River showed that not only can natural 'native' sources be the major contributors to P in Australian waterways, but that the P concentration on the fine sediment provides sufficient bioavailable P for algal bloom formation (Murray 1996; CSIRO 1996). Martin (1996) has also demonstrated that most of the total P reaching the Wyong River comes in association with sediments derived from channel bank erosion.

Evidence from this study and other recent studies suggests that the principles underpinning the conventional view that P is primarily derived from superphosphate fertilisers and sewage/industrial effluents are not entirely valid for many turbid Australian inland waterways. It is recommended, therefore, that other rural catchments be investigated to further test the hypothesis that native P is a major contributor to P budgets in turbid waterways. If this hypothesis is confirmed, one can argue that the approaches currently used to control eutrophication and algal blooms may not be effective in some catchments and, therefore, alternative strategies must be considered.

Since the formation of algal blooms depends on a complex interaction of a range of biophysical processes, it is dangerous to assume that the causes and factors affecting algal growth are the same for all catchments. Different catchments may have different sources of nutrients and different limiting factors, and thus different strategies must be adopted to manage the algal blooms problem. In some catchments the conventional view about sources and transport of P may be valid, in other catchments N rather P could be the main limiting nutrient (Grace *et al.* 1997), and in other catchments flow or the ecological balance of the aquatic ecosystem could be the critical factor to control algal growth. Consequently, more scientific work needs to be carried out to better understand the physical, chemical and ecological dynamics that cause algal growth. It is also important to note that sustainable solutions to algal blooms require the application of an integrated management approach that takes into consideration both the biophyscial processes and socio-economic factors. Since the biophysical processes control the inherent characteristics and constraints of land and water resources (Al Bakri *et al.* 1998), scientific investigations are essential precursors for the development of sustainable management strategies to control algal blooms. Indeed, unless we get the scientific rationale of the problem right it is unlikely that sustainable management of algal blooms can be achieved.

Collaboration between scientists, managers and farmers is imperative to develop sustainable management systems to combat algal blooms. Science can play an important part in the effort to deal with these problems, but scientists should work harder to communicate their findings to the wider community, be responsive to the needs of the managers and farmers, and ensure that scientific knowledge is an integral part of the management process. On the other hand, the rural community should encourage and support more scientific research to improve our regional understanding of the nutrient-sediment interaction, ecological dynamics and catchment behaviour. This understanding is fundamental to establishing a sound basis for the development of cost-effective management options to reduce the frequency of algal blooms in rural Australia.

References

Al Bakri, D., Chowdhury, M., and Reddan, B. (1995). *Assessment of water quality of Orange water supply catchment.* The 3rd Annual Conference of Soil and Water, 12–15 September 1995, Sydney.

Al Bakri, D., and Chowdhury, M. (1997). *Water quality of Orange water supply catchment: Physico-chemical properties and nutrients.* Workshop on Water Quality in Orange, Orange Agricultural College, The University of Sydney, 23 May 1997, Orange.

Al Bakri D., and Chowdhury M. (1998). *Biophysical barriers to sustainable water resource management: an Australian perspective.* International Workshop on Barriers to Sustainable Management of Water Quantity and Quality, 12-15 May 1998, Wuhan, China.

Australian and New Zealand Environment and Conservation Council (1992). *Australian water quality guidelines for fresh and marine water.* (Australian and New Zealand Environment and Conservation Council: Melbourne.)

Beck, P., and Long, G. (1994). *Lake Forbes Water Quality Management Plan.* (Department of Water Resources: Forbes, NSW.)

Blue-Green Algae Task Force (1992). *Blue-green algae.* Final Report of the Task Force, NSW Department of Water Resources, Sydney.

Buchan, A., and Diacono, B. (1995). *Water quality in the Murrumbidgee: What the community wants.* (Department of Land and Water Conservation: New South Wales.)

Caitcheon, G., Donnelly, T., and Wallbrink, P. (1995). Nutrient and sediment sources in Chaffey Reservoir Catchment. *Australian Journal of Soil and Water Conservation* **18**(2), 412–419.

Carson, C. (1992). What can be done about toxic algal blooms? *ECOS* **72**, 14–19.

Causely, I. (1989). Water quality in New South Wales: meeting the challenge. *Quarry Mine and Construction News* **28**(9), 11.

Chowdhury, M., and Al Bakri, D. (1998). Phosphorus inputs to a reservoir in Orange, NSW. *Water* July/August 1998.

Cook, P. (1983). World availability of phosphorus: an Australian perspective. In *Phosphorus in Australia*, CRES Monograph 8. (Eds A.B. Costin, and C.H. Williams.) (Centre for Resource and Environmental Studies, Australian National University: Canberra.)

Cosser, P. R. (1989). Nutrient concentration–flow relationships and loads in the South Pine River, South-Eastern Queensland. *Australian Journal of Marine Freshwater Research* **40**, 613–630.

CSIRO (1994). *Nutrients and sediments. Sources in Chaffey Reservoir catchment.* (Division of Water Resources, CSIRO: Canberra.)

CSIRO (1996). *Sources of suspended sediment and phosphorus to the Murrumbidgee River.* Division of Water Resources, Consultancy Report No. 95-32. (CSIRO: Canberra.)

Cullen, P. (1986). Managing nutrients in aquatic systems: the eutrophication problem. In *Limnology in Australia.* (Eds P. Dekker, D. Williams, and W. Junk.) (CSIRO: Melbourne.) pp. 540–554.

Cullen, P. (1995). Managing non-point sources of phosphorus from rural areas. *Water Journal* **22**, 12–14.

Cullen, P., and O'Loughlin, E.M. (1982). Non-point sources of pollution. In *Prediction in Water Quality.* (Eds E.M. O'Loughlin, and P. Cullen.) (Australian Academy of Science: Canberra.)

Donnelly, T.H., Olley, J.M., Murray, A., Caitcheon, G., Olive, L., and Wallbrink, P. (1996). Phosphorus sources and algal blooms in Australian catchments. In *Proceedings of the National Conference on Downstream Effects of Land Use*, Rockhampton, 26–28 April 1995. (Eds H.M. Hunter, A.G. Eyles, and G.E. Rayment.) (Department of Natural Resources, Brisbane.)

Grace, M.R., Hart, B.H., Oliver, R.L., and Rees, C.M. (1997). Algal blooms in the Darling River — are phosphates to blame? *Chemistry in Australia* August 1997, 2–5.

Gonsiorczyk, T., Casper, P., and Koschel, R. (1997). Variations of phosphorus release from sediments in stratified lakes. *Water, Air and Soil Pollution* **99**, 427–434.

HACH (1992). *Water analysis handbook*, 2nd edn. (HACH Company: Colorado, USA.)

Harris, G. (1995). Eutrophication — are Australian waters different from those overseas? *Water Journal*, **22**, 9–12.

Harris, G.P. (1994). *Nutrient loadings and algal blooms in Australian waters —a discussion paper.* Occasional Paper No. 12/94. (Land and Water Resources Research and Development Corporation: Canberra.)

Hart, B.T. (1996). Sediment–nutrient uptake and release: Progress towards predictive models. *National Workshop on Sediment–Nutrient Interaction.* 30 April, 1996, Albury.

Hecky, R.E., and Kilham, P. (1988). Nutrient limitation of phytoplankton in freshwater and marine environments: a review of recent evidence on the effects of enrichment. *Limnology and Oceanography* **33**, 796–822.

Kilham, P., and Kilham, S.S. (1990). Endless summer: internal loading processes dominate nutrient cycling in tropical lakes. *Freshwater Biology* **23**, 379–389.

Kovac, M., and Lawrie, J.A. (1990). *Soil landscapes of the Bathurst area 1:250000 Sheet.* (Soil Conservation Service of New South Wales: Sydney.)

Martin, P.M. (1996). *Slumping of phosphorus rich bank sediments as a significant source of total phosphorus loading in the Wyong River, Central Coast, New South Wales.* (University of Sydney: Sydney.)

Murray, A. (1996). Transport of phosphorus in turbid Australian rivers: An alternative hypothesis. *National Workshop on Sediment–Nutrient Interaction.* 30 April 1996, Albury.

O'Connor, R. (1992). *Water quality of Kedumba River system 1989–1990.* (NSW Water Board: Sydney.)

Oliver, R., Hart, B., Douglas, G., and Beckett, R. (1993). Phosphorus speciation in the Murray and Darling Rivers water. *Journal of the Australian Water and Wastewater Association* **20**(4), 24–26.

Robinson, G., and Hatfield, E. (1992). Application of flow duration curves in calculating nutrient/pollution loadings. Poster paper presented at *Environmental Biometrics Conference, Sydney, Australia, 14–15 December 1992.*

Scott, M., Warren, A., Meakin, S., and Watkins, S. (1994). Orange geology 1:100,000 Basement Geology Series. *Field Geology Conference*, October 1994, Orange.

Shafron, M. (1995). Blooming algae. *Agricultural Science* **8**(1), 44–47.

Taylor, S. (1994). *Macquarie River catchment: Land management proposals for the integrated treatment and prevention of land degradation.* (CaLM: Bathurst.)

Verhoeven, J. (1993). Implementation of the New South Wales algal management strategy. *Australian Journal of Soil and Water Conservation* **6**(3), 30–34.

Wasson, R.J., Donnelly, T.H., and Murray, A.S. (1996). Imports can be dangerous — Appropriate approaches to Australian rivers and catchments. In *Proceedings of the First National Conference on Stream Management in Australia.* (Eds I. Rutherford, and M. Walker) pp. 313–319. (CRC Catchment Hydrology, Monash University: Melbourne.)

Wyborn D. (1994). Regional metamorphism in the Bathurst 1:250,000 sheet area. *Field Geology Conference*, October, 1994, Orange, NSW.

Beefing up our trade: health and environmental concerns and rural exports

Marie Wynter

Introduction

The future of Australian rural communities is intimately entwined with the forces of globalisation and the expansion of international trade. As competition for markets becomes more intense, a trend towards the politicisation of environmental and consumer health and safety standards is emerging. This is due in part to a surge in consumer demand for higher environmental and health and safety standards, yet we can also see an increasing use by protectionist groups of those same standards to slow international trade and further their own agendas.

It is likely that environmental concerns will become increasingly important as this trend continues. This chapter examines a dispute involving the use of hormones in beef production (which I will call 'the hormone dispute') recently adjudicated by the World Trade Organization (WTO). It provides an illustration of the manner in which the dramatic politicisation of environmental and consumer standards has the potential to upset the livelihood of rural communities and severely jeopardise international trading relations.

A significant factor influencing the outcome of that case is the new agreement on Sanitary and Phytosanitary Standards ('the SPS Agreement'), signed by most of the world's trading community as part of the Uruguay Round agreements. This chapter argues that whilst the SPS Agreement will make it harder for nations to erect non-tariff trade barriers to support environmental or consumer agendas, these barriers will continue to be erected in the future.

This chapter first describes the background to this dispute and its effects on the major stakeholders. It then analyses the interpretation of the SPS Agreement within the context of the hormone dispute, and finds that the hormone ban is but a sign of things to come. This chapter suggests that Australian rural producers need to think strategically and be proactive about addressing environmental concerns. This will facilitate capture of the growing market for sustainable produce and avoid the negative trade impacts of green protectionism.

The European ban on the use of growth promoting hormones on beef production

The hormone ban has been one of the most significantly disputed non-tariff trade barriers dogging agricultural exports. The European Communities' continuance of the hormone ban was challenged by the United States and Canada at the supranational level, giving the WTO its first opportunity to rule on the interpretation and the application of the SPS Agreement.

The dispute raises fundamental questions about the ability of countries to enact legislation to address the concerns of their citizens and which has the citizens' popular support. If countries want to enjoy the benefits of being part of a global community, they have to accept the inevitable loss of sovereignty which this involves. The extent to which this is true in the area of environmental protection and consumer health and safety has been influenced by the outcome of the hormone dispute, the resulting jurisprudence clarifying some of the ground rules for the ability of member countries to develop laws which have an effect on trade.

Background to the dispute

From Italy in 1980 reports of babies developing secondary sexual characteristics, such as growing breasts and menstruating came to the public attention (see eg Froman 1989; Halpern 1989). These reports confirmed the worst fears of consumers worried about the use of additives, including growth promoting hormones, in livestock production. The growth abnormalities were attributed to high levels of diethylstilbenes (DES), a synthetic growth hormone, which was found to be present in veal extract baby food (United States Department of Agriculture 1987; Froman 1989); however a definitive link between the presence of the hormone and the abnormal growth was never scientifically proven (Froman 1989; Vogel 1995). It is important to note that the DES found had been used illegally, and probably administered incorrectly. It was a clear case of hormone abuse.

Until the late 1970s, DES was used in the treatment of pregnant women and was used as a livestock feed additive (Froman 1989). Following demonstration of the carcinogenic effects of DES in both humans and laboratory animals, the use of DES was gradually banned in Europe and the United States (Vogel 1995). The public was however becoming increasingly concerned about the increasing use of illicit DES and the Italian reports prompted widespread boycotts on veal in Europe and strong calls for further bans on hormones in the beef industry. The boycotts had a considerable effect upon the sale of beef; for example the French Federal Consumers Union Boycott led to a 70% decline in purchases (Vogel 1995; Report of the Panel on European Communities; Report of the Appellate Body on European Communities). See also Colbourn, Dumanoski and Myers (1996) for details on the effects of DES on human health and fertility.

The concern and outrage exhibited by the public, as well as numerous political recommendations, led Europe's Council of Ministers to direct a total ban on the use of all growth hormones. On 16 March 1988 a valid directive was passed, banning all hormones other than as treatment for 'management breeding purposes'. Countries wishing to export to Europe had to accordingly modify the manner in which they produced their beef.

Politics clearly influenced the adoption of a total ban. Consumer pressure, particularly in Italy, Belgium and France, was enough to raise hormone tolerance standards in countries within the community, but those standards were not uniform throughout the community. The total ban removed the disruption to trade caused by the different standards, and the need for extensive border controls to ensure hormone-treated beef was not imported from elsewhere (Vogel 1995).

Reactions to the ban

Australia responded to the European hormone ban by instituting a control system which satisfied the European criteria, presently estimated to cost the Australian meat industry about $10 million a year (Brenchley 1997, p. 14). Accordingly, Australia has retained its exports to the European Community, now worth $83 million per annum (Brenchley 1997). Although growth promoting hormones can act as a significant economic aid to production, Australian beef production is primarily grass-fed, for which hormones, whilst representing some cost savings, are not integral. Europe is a strategically beneficial market as it is regarded as a 'benchmark' market in terms of meat quality. Maintaining supply to Europe signals to the world market that Australian beef is of a high standard and is safe to eat. Australia considers that no health and safety, domestic demand, or economic considerations warrant a total ban, and continues to allow producers supplying home and other markets to use growth promoting hormones. Fortunately, our largest export markets are either significant users of the hormones (the US) or have not indicated they consider hormone use a risk (eg Japan).

Despite the inconvenience and expense caused by the administration of, and compliance with, the European hormone ban, producers have now adjusted to the effects of the ban. Partly because of the hormone restrictions, Australian producers are able to sell beef at a premium into the European market. Indeed, although some Australian users of growth promoting hormones are renewing their licences, quite a few lot feeders have discontinued their use of hormones, finding that they do not provide the advertised financial benefits, that their use locks them out of markets, or that they prefer to produce in a manner which avoids consumers' negative perceptions of hormone use.

The United States struggled hard against the hormone ban, engaging in extensive negotiations and consultations with Europe to have it removed. The beef hormone dispute even appeared to threaten the Uruguay Round negotiations as it was seen as symptomatic of the larger differences which the US and the EC had over support to the agricultural sector (Froman 1989). It is believed that the US proposed strengthening the GATT rules on sanitary and phytosanitary standards as a way out of the hormone impasse (Vogel 1995; Steinberg 1997). When the ban was implemented, the US invoked US$1 billion worth of punitive sanctions on EC goods to ameliorate its loss of trade. It subsequently brought an action against the EC at the WTO to abolish the ban on hormones which it argued was unscientific and out of line with international standards, guidelines and recommendations; Canada has brought an essentially identical action. The US–EC panel was convened on 20 May 1996 and the Canada–EC panel was convened on 16 October 1996.

Given the overall minor impact of the hormone ban on Australia, we reserved third-party rights to make submissions to the WTO dispute resolution panel on the hormone ban, but did not join as a plaintiff in the actions taken by the US or Canada against the EC. New Zealand and Norway also reserved third-party rights to contribute to the submissions.

The SPS Agreement and the hormone rulings

The WTO issued two essentially identical first instance panel reports ruling against the European ban on 18 August 1997. These were appealed. On 16 January 1998 the Appellate Body released a report which was very critical of the panel's findings in their reports. Although it confirmed the essence of the panel's finding that the present European ban is contrary to the rules of the SPS Agreement, it has left the way open for a more finely tailored ban to be put in place. Therefore it is not yet time to start using hormones in beef destined for Europe.

The hormone rulings are very important for a number of reasons. This was the first opportunity for the World Trade Organization (WTO) dispute settlement body to consider and clarify the application of the SPS Agreement within the context of the Uruguay Round Agreements as a whole. Therefore it establishes how the SPS Agreement interrelates with the other Agreements, the burden of proof necessary to bring a case and defend a standard, the extent to which the WTO will scrutinise a standard and how experts may contribute to that process, and what is expected of a member in terms of justifying a standard.

Importantly, the WTO has clarified that the SPS Agreement is a *free-standing* agreement and no *prima facie* case of discrimination or different treatment need be shown for an SPS Agreement investigation to take place as is necessary under the General Agreement on Tariffs and Trade (GATT 1994). Instead, a complainant needs to show that *prima facie* the member has chosen SPS standards which are higher than the relevant international standards and have been put in place without a risk assessment having been done regarding the threat which the standards intend to protect against.

The burden of proof

Once a complainant has shown that the member's standards are higher than the relevant international standards, guidelines or recommendations, and that no adequate risk assessment has been done, or that the standards have been applied in some other manner which is inconsistent with the SPS Agreement or the GATT 1994, the burden shifts to the member to prove that nevertheless the standard does comply with the SPS Agreement. This distribution of the burden of proof is consistent with the manner in which proof is treated in other WTO Agreements. See in particular *United States — Measure affecting imports of woven wool shirts and blouses from India* and also *United States — Restrictions on imports of cotton and man-made fibre underwear*. (Report of the Appellate Body on United States.)

The Appellate Body's discussion on this point, including its extensive criticisms of the panel's approach, serves as a timely reminder that both the structure and words of the Agreement must be carefully observed so as to preserve the rights of the parties and the essence of the Agreement as a whole. The negotiators of the Agreement attempted to balance the democratic right of nations to choose their own level of protection against a means of ensuring that the chosen protection reflects genuine health and environmental concerns and not protectionist concerns. Observing the words of the Agreement allows predictability in ensuring that this objective is fulfilled.

Regard for International standards, guidelines and recommendations

The US and Canada were able to trigger the application of the SPS Agreement to the European hormone ban by showing, *prima facie*, that the standards applied by the EC with respect to the natural hormones oestradiol–17β, progesterone and testosterone, and the synthetic hormones zeranol and trenbolone, were higher than the international standards set by the Codex Alimentarius ('the Codex') for those substances. They also provided sufficient evidence to show that, while risk assessments had been done by the EC, international bodies and independent academics, the EC standards were not based on those risk assessments but on the possibility of risk which had not been assessed.

The SPS Agreement uses international standards as a benchmark by which to achieve the progressive harmonisation of SPS standards implemented by member countries. Harmonisation is desirable as not only does it reduce the complexity of legislation which importers have to deal with (in itself an impediment to trade), but minimises the use of such standards for protectionist purposes.

National SPS standards may either be based on international standards (Article 3.1); *conform to* international standards (Article 3.2) or be *higher than* international standards if there is scientific justification or a member determines it to be appropriate (Article 3.3).

Where the chosen standard conforms to — that is, is the same as — the international standards, it is rebuttably presumed to be consistent with the SPS Agreement and the GATT 1994. Members have an autonomous right to choose a level of protection which is higher than that afforded by international standards, as emphasised by the Appellate Body in its report, and may either base their chosen level of protection on the international standards or disregard the international standards. This right is qualified in both situations by the SPS Agreement requiring that members justify their ban by basing their standards on a risk assessment according to Article 5.1. This is a balancing mechanism intended to ensure health and environmental standards can be used, but not abused.

The EC unsuccessfully argued that the international standards established by the Codex did not apply to this dispute. It argued that whereas the EC standards affect the use of the hormones, the Codex does not set standards for the use of hormones but rather *maximum residue limits*. The panel disposed of this argument by stating that the standards set by both the EC and the Codex reflect a *level of protection*, of which the EC's level is considerably higher. The panel then went into detail as to how Article 3 should be read within the context of the rest of the agreement.

The EC appealed this part of the panel's judgement. Although the Appellate Body agreed that the Codex standards were apposite to the dispute, it was strongly critical of how the panel interpreted Articles 3.1 and 3.3 and their interrelation with Article 5. It pointed to a number of errors of law made by the panel, only some of which did it attempt to amend. For example, the panel held that for a member's standards to be in compliance with Article 3.1, they must be the same as — that is, conform to — international standards, guidelines or recommendations. The Appellate Body pointed out that this is incorrect. Only Article 3.2 requires standards to conform to international standards, guidelines or recommendations. Article 3.1 merely requires standards to be based on international standards, guidelines or recommendations, where they exist.

It is interesting to note that the SPS Agreement has partially reified international standards. As the Appellate Body took pains to point out, members do not have to unthinkingly implement relevant international standards to be in compliance with the SPS Agreement, but may choose national standards adapted to their own national conditions — including the anxieties of their consumers. Nevertheless, the congruence between a member's standards and international standards acts as a trigger for whether an action taken under the SPS Agreement can be brought.

This adds considerably to the weight of international standards, guidelines and recommendations such as those provided by the Codex. They have always been recommendations, but now have gone beyond the point of being merely hortatory and only *potentially* evidence of customary practice, to be standards which members ignore at their peril. This will undoubtedly further politicise standard making at the international level. Whether the effect will be to achieve the upwards or downward harmonisation of international environmental, consumer health and safety standards will have to be assessed over time.

Setting the appropriate level of sanitary of phytosanitary protection

Article 5.1 states:

> 'Members shall ensure that their sanitary or phytosanitary measures are based on an assessment, as appropriate to the circumstances, of the risks to human, animal or plant life or health, taking into account risk assessment techniques developed by the relevant international organizations.'

As stated above, a risk assessment is the necessary foundation for all national SPS standards, unless the standards conform to international standards (Article 3.2). The risk assessment may be performed by the member or by other members, or by an international organisation. Such an assessment helps legitimise a member's choice of their appropriate level of protection.

Risk assessment involves both a scientific determination of the uncertainty of an event and any ensuing damage occurring, and a psychological evaluation of the attitudes people have towards that event. The process of risk assessment is inherently a subjective one and involves interconnecting issues including the concern which individuals, groups or cultures have towards an event; social issues of fairness and ethics such as whether the risk is voluntary, involuntary or unknown; and political and economic considerations such as whether the benefits outweigh the risk and whether it is cheaper to live with the risk than remove it (Burton *et al.*1978; Handmer *et al.* 1991; Brown 1992).

In assessing whether the EC had properly conducted a risk assessment, the panel noted that 'an assessment of risks is, at least for risks to human life or health, a *scientific* examination of data and factual studies; it is not a policy exercise involving social value judgments made by political bodies' (para. 8.94 of the Hormone Panel Report, *United States — EC Measures Concerning Meat and Meat Products (Hormones)*; para. 9.97 of the *Hormone Panel Report, Canada — EC Measures Concerning Meat and Meat Products (Hormones)*). Therefore it clearly distinguished the scientific and subjective elements of the determination, finding that subjective considerations could not override scientific considerations. It characterised the scientific aspect of the examination as *risk assessment*, and the remaining investigation as *risk management*.

Rather than becoming involved in an actual investigation of the science supporting the EC ban, the panel held that it was for the defendant to establish whether its measures were based on a risk assessment. The panel could not undertake a *de novo* examination of the science. It did however hear extensive evidence from six scientific and technical experts in order to gain a more scientific appreciation of the effects of the hormones under dispute and the evidence presented by the parties. This was a sensible move and adds credibility to the judgement.

Past criticisms of GATT/WTO panels have been that they have failed to consult scientific or technical experts when faced with rulings on measures of environmental and health-related matters (see McDonald 1997). WTO adjudicators' expertise lies predominantly in trade and international law, not science, and critics have been sceptical of their qualifications to make decisions on complex technical matters as well as their familiarity with, and acceptance of, paradigmatic views in which trade is not central. These concerns remain relevant, to some extent, to this dispute as, while experts were consulted, this was on an individual basis. It was the panel which was forced to weigh and reconcile the disputed scientific evidence, possibly without some of the scientific and methodological rigour which a panel of scientific experts, working together to provide a cohesive report, would have been able to contribute. Perhaps it was felt that it would be too difficult to reach a consensus amongst the experts, or that such an exercise would unduly focus on the scientific elements of the risk assessment, excluding the subjective considerations which the EC was in the best position to assess.

The European Commission had ordered two major scientific inquiries into the health effects of hormones used for growth promotion purposes: the Lamming and Maddox reports. Both inquiries found that *properly administered* growth hormones do not pose a risk to human health. The most recent inquiry in 1995 suggested further research into the metabolic effects of hormones used in combination. The commission also conducted two other inquiries, published in the Pimenta and the Collins reports, which found that there was a high potential for *hormone abuse* to occur unless a total ban on hormones used for growth promotion purposes was maintained. Concerns regarding this potential for abuse outweighed the scientific evidence regarding the low health risk associated with properly administered hormones, and notwithstanding the results of the 1995 inquiry, the European Parliament voted 366 to 0 in support of a resolution to maintain the ban. There were 260 parliamentarians who did not attend the vote.

The panel accepted these inquires, as well as studies done by JECFA (an arm of the Codex Alimentarius) and other academic articles, as risk assessments done in accordance with Article 5 for all of the hormones in dispute except MGA (melengestrol acetate). The panel found however that the EC had met neither the minimum procedural requirements of Article 5.1 (to take into account the conclusions of the reports) nor the substantive requirement of that article (to ensure that the conclusions of the risk assessment were reflected in the chosen measure). The ban on the use of MGA had been adopted without a risk assessment and therefore was inconsistent with Article 5.1.

The Appellate Body was strongly critical of the panel's interpretation of Article 5.1. Firstly it criticised the panel's separation of the investigation into risk assessment and risk management. It found that the term 'risk management' had no basis in the text, and that such a separation led to an unduly restrictive approach to risk assessment.

Second it criticised the panel's exposition of a minimum procedural requirement that a defendant must show that they 'took into account' the conclusions of the risk assessment. It pointed out that there is no 'minimum procedural requirement' set out in the text of the SPS Agreement and the test set by the panel was in any event unhelpful. According to the Appellate Body, the panel's interpretation would allow studies to be taken into account and nevertheless ignored. The Appellate Body stressed that there must be an *objective* and substantive relationship between the risk assessment and the standards adopted, that is 'the results of the risk assessment must sufficiently warrant—that is to say, reasonably support—the SPS measure at stake ... there [must] be a rational relationship between the measure and the risk assessment' (para. 193, Appellate Body Report).

Finally, the Appellate Body identified for the EC the means by which it could re-tailor its ban so as to comply with the SPS Agreement. The Appellate Body pointed out that whilst detailed studies had been presented to the panel discussing the health effects of hormones applied according to good veterinary practice, the details of the studies which examined the risks associated with hormone abuse — for example those made in conjunction with the Pimenta Report — were *not* put before the panel. Demonstrating an objective relationship between the risk assessment and the national standards currently in place would require the results of such a study. The Appellate Body accepted that the political decision to ban the hormones appeared to rest on the risk to humans if the hormones were abused, but was unable to find that this was based upon a risk assessment as required by Article 5.1.

The Appellate Body pointed out that it was entirely appropriate for risk assessments to take into account the circumstances of how a product was produced, whether the hormones could have been properly or improperly administered, the economic incentives for doing so, and so on. It also held that a risk assessment does not have to arrive at a 'monolithic' conclusion, but may represent both 'mainstream' and 'divergent' scientific opinions. Moreover, it could be entirely within the terms of the SPS Agreement for a responsible and representative government to choose to base its SPS standards on the body of divergent opinions 'coming from qualified and respected sources'. Reflecting 'divergent' opinion in SPS standards was especially reasonable 'where the risk involved is life-threatening in character and is perceived to constitute a clear and imminent threat to public health and safety' (para. 194, Appellate Body Report).

It should be noted that language of 'clear and imminent threat' is not to be found in the SPS Agreement, nor is it sensible to limit the acceptance of divergent science to these situations only. A risk which is sizeable yet distant because of the cumulative effects of factors contributing to the risk should not be marginalised merely because it is supported by qualified and respected albeit 'divergent' opinion. This would be to ignore the manner in which scientific opinion is constructed. Rather than science being a slow accretion of facts leading to inevitable conclusions with whom all parties are in accord, science often proceeds by 'lurches' when unexpected discoveries and compelling syntheses promote a revision of thinking such that an area previously regarded as a minority position becomes the new mainstream. Take for example Rachel Carson's publication of *Silent spring* which heralded a shift in attitude towards the incremental effects of pesticides on human health and the environment (Carson 1963). It is important that nations be allowed to implement SPS measures from the time in which the minority opinion gains sufficient credibility that the mainstream is starting to take notice of the scientific concerns raised, regardless of whether there has been a landslide movement to embrace that opinion.

Having said that the complete eradication of risks is a worthy goal to strive for, yet is ultimately unrealistic. The central question for society remains not whether we want to live with risk, but what sorts of risks we do want to live with, and how we are to prepare to deal with the risks when the inevitable eventuates (Schroeder 1986; Gillette and Krier 1990; Handmer *et al.* 1991).

The WTO panel has provided important insights into how these elements are to be balanced in the future. Members do not have a free hand to implement any standard without it being backed up by a thorough risk assessment. The WTO will not accept the risk assessment unquestioned, but will seek expert opinion to inform it of the current strands of scientific opinion on the issue. Members may implement standards which reflect divergent scientific opinion if that opinion is from a sufficiently qualified and respectable source; however they must show sufficient connection between the risk assessment and the standards in place. The problem with the European ban was there was no such connection.

Consistency of levels of protection and resulting discrimination or restriction on international trade

One of the goals of the SPS Agreement is to encourage members to maintain standards which are internally consistent. This is a goal, not a legal obligation. In order to promote that goal, Article 5.5 sets out the requirement that:

> '... Each member shall avoid arbitrary or unjustifiable distinctions in the levels it considers to be appropriate in different situations, if such distinctions result in discrimination or a disguised restriction on international trade.'

Furthermore, Article 5.5 must be read in context with Article 2.3 which points out:

'Members shall ensure that their sanitary and phytosanitary measures do not arbitrarily or unjustifiably discriminate between Members where identical or similar conditions prevail, including between their own territory and that of other Members. Sanitary and phytosanitary measures shall not be applied in a manner which would constitute a disguised restriction on international trade.'

Article 5.5 consists of three elements, set out by the Appellate Body at paragraph 214 of its report:

1   the member … has adopted its own levels of appropriate protection against risks to human life or health in several different situations;

2   the levels of protection exhibit arbitrary or unjustifiable differences ('distinctions' in the language of Article 5.5) in their treatment of different situations; and

3   the arbitrary or unjustifiable differences result in discrimination or a disguised restriction of international trade.

To compare levels of protection existing in different situations, situations where the same substance or the same health effects are involved may be compared. The EC had adopted differing levels of protection in respect of:

* natural hormones used for growth promotion purposes;

* natural hormones occurring endogenously in meat and other foods;

* natural hormones used for therapeutic or zootechnical purposes;

* synthetic hormones when used for growth promotion purposes; and

* the use of carbadox and olaquindox, anti-microbials which promote growth in piglets, and are known carcinogens.

The distinction between the levels of protection for natural hormones present endogenously in meat and other foods and natural hormones which were added for growth promotion purposes was found to be justifiable by the Appellate Body, overturning the panel's ruling. Differences in levels of protection between natural hormones used for growth promotion purposes and for therapeutic or zootechnical purposes were also considered justifiable by the Appellate Body. Differences between the levels of protection regarding natural and synthetic hormones used for growth promotion purposes and carbadox and olaquindox were, on the other hand, found to be *unjustifiable* in the sense of Article 5.5. Despite being known carcinogens, the standards for carbadox and olaquindox were set considerably lower than the hormone standards.

This last finding did not mean that the measures were incompatible with Article 5.5. The Appellate Body stressed that all three elements of that Article had to be satisfied for a contravention to have occurred. In this respect the Appellate Body *overruled* the panel's finding that the distinctions in levels of protection resulted in discrimination or a disguised restriction on international trade. It firstly overruled the manner in which the panel applied existing WTO law to the interpretation of this section of the Article, and then overruled the inference the panel derived from the documents preceding and accompanying the enactments of the EC Hormone Directives that the EC had protectionist motivations for enacting a ban on total hormone use. The Appellate Body stated that these documents and other evidence before the panel:

'make clear the depth and extent of the anxieties experienced within the European Communities concerning the results of the general scientific studies (showing the carcinogenicity of hormones), the

> *dangers of abuse (highlighted by scandals relating to black-marketing and smuggling of prohibited veterinary drugs in the European Communities) of hormones and other substances used for growth promotion and the intense concern of consumers within the European Communities over the quality and drug-free character of meat available in its internal market.'*

Therefore it appears that once the EC completes a proper risk assessment of the actual risks of hormone abuse occurring somewhere in the meat production chain, and the possible impacts of hormone abuse on human health, they may re-enact hormone standards which sensibly reflect that risk. These standards will have a good chance of being non-discriminatory, even if they are set high and do not reflect the same level of protection as that pertaining to carbadox and olaquindox.

The Appellate Body ruling has consequently left a debate waging between the US and the EC as to who has actually won this case. Whereas the EC is taking the position that it may keep the present ban in place until it completes the risk assessment procedures again, the US and Canada state that Europe must remove the ban immediately.

The ruling also brings to light the possible need for the EC to examine whether the carbadox and olaquindox standards are actually set high enough. The SPS Agreement does not however include a mechanism for raising standards, only assessing whether they are too high.

A final point to make is that it is quite difficult to detect hormone abuse, thus throwing into doubt the ability of the hormone ban to be properly enforced. Using hormones for growth promoting purposes does appear to give, in some cases, significant financial rewards, causing some producers to use not only the banned hormones discussed above, but also more dangerous drugs such as clenbuterol. Clenbuterol is apparently responsible for the hospitalisation of 135 people in Spain after they ate contaminated meat, and for the death of farm workers in Ireland when they mixed it with animal feed. Suggestions indicate that Spain, Italy and Belgium have a very high incidence of illegal hormone use, and that Germany has a black market in veterinary drugs, worth $150 million. This appears to support statements made by Dr Lamming after the EC adopted a total ban on hormones that '[t]he European Economic Community's ban on hormone use in calves will collapse because it is unscientific, impossible to police, and because it will lead to a huge black market across Europe in cattle implants.' (USDA 1987, at p. 17, quoting 'Hormone ban by EC seen unlikely to be reversed' *Food Chemical News* (6 October 1986), p. 6). Following the initial report made by Lamming as to the low health risks associated with properly administered hormones, the EC cancelled the Lamming working group and implemented a strict hormone ban. This was contrary to internal and external science, but in line with consumer concerns.

The widespread abuse of these hormones indicates that consumers are right to be concerned about the health risks involved, although distinguishing between the health effects of hormones applied properly and those misused is important. It also emphasises that the political reaction has failed to adequately address the source of that concern, and that mechanisms other than a blanket ban are necessary. One suggestion for resolving this issue is to allow consumers to choose whether they are willing to eat hormone-treated meat. The US has indicated that it will not tolerate labelling requirements which state the meat is 'hormone treated'. It is unlikely that meat labelled in this way would be attractive to consumers, and producers already engaging in illegal hormone use would be unlikely to comply with this requirement as the consumers would provide them with little incentive.

The hormone dispute — a sign of things to come?

The Uruguay Round saw the linking of a number of 'new' issues to negotiations on international trade: how trade and trade standards can be used to conserve environmental resources and contain environmental damage; raise the global level of consumer health and safety; and raise living and working conditions and encourage global observance of basic human rights.

There is considerable concern within the trading community that the linking of these issues to international trade concerns will proliferate, possibly overwhelming the liberalisation gains which have been so hard won since the GATT was first agreed in 1947 (Grimwade 1996).

It is unlikely that environmental concerns will abate in the face of continuing reports of environmental damage around the world, and as trade measures are seen as one of the few mechanisms to actually coerce compliance with environmental standards, attempts to use them will continue.

Thus the hormone dispute is a sign of things to come, and also provides a few salutary lessons for Australian rural producers.

As the analysis of the SPS Agreement above has indicated, nations can only implement SPS standards which are scientifically based and are backed with a properly prepared risk assessment. Once this is done however, the standard, even though it is out of line with international practice, can be maintained.

Also, countries will tend to keep trade barriers in place while they can get away with it. The WTO does not prevent countries raising trade barriers; it merely allows action to be taken against members with trade barriers proven to be GATT-illegal. In the case of the hormone ban, this has taken almost 10 years and it is still not over. The EC will undoubtedly redo its risk assessments taking into account the comments by the Appellate Body, and in the mean time are likely to keep its bans in place.

In light of this, Australia should start preparing for countries placing increased importance on the environmental impacts of goods produced, and the consumer health and safety impacts of the goods themselves. Sustainable agricultural practices can be seen as a way to ensure a long-term export market as well as long-term production capabilities.

Products which are grown, produced or harvested using methods of comparatively lower environmental impact will become increasingly competitive as environmental health and safety concerns rise. Producers who are at the forefront of developing such products will be in a better position to avoid market entry being denied, and will also be in a better position to educate their markets so as to maintain their relationship and pave the way towards expansion.

With comparatively clean air and soil, and rural production largely occurring away from areas with a concentration of heavy industry, Australian agricultural products have a better chance of being cleaner than those of other countries. The uptake and continuance of Landcare will work towards reducing and ameliorating present environmental concerns in agricultural production, as will technological advances.

Europe is one of Australia's largest trading partners, and demands high environmental standards for agricultural as well as other goods. Asia is a natural and growing market for much of Australian produce. Already Japan demands high consumer health and safety standards which are likely to extend towards environmental concerns as these issues grow in importance. This trend is likely to be followed by the rest of Asia, albeit at a significantly slower rate. As emphasised by a number of trade economists, as nations become affluent, increasing attention is paid to environmental concerns, and more money is spent on ameliorating environmental harms and purchasing environmentally 'friendly' products (see eg Anderson and Blakehurst 1992). The market for 'clean and green' products is unlikely to be a 'passing fad' and is likely to grow significantly; fortunately we do have some time for strategic preparations.

Finally, while the costs for conversion to more sustainable agricultural production can act as a substantial disincentive, modifications under the Uruguay Round allow governments, in limited circumstances, to make payments to producers to meet the costs of complying with higher environmental standards. It is important that producers be aware of this provision so that they can lobby for support when and if these costs are incurred (Article 8. para. 8.2(c) of the WTO Agreement on Subsidies and Countervailing Measures; see Wynter 1996).

It is not in our interests to knock down other nations' health and environmental standards unless they are actually unfair. Given our unique and often vulnerable environment, Australia could easily fall victim to potentially disastrous exotic plants, pests and diseases if we too aggressively sought a restriction in WTO law allowing countries to maintain latitude in choosing their appropriate level of protection. For example, Australia's ban on the import of salmon from Canada is currently under scrutiny at the WTO, our ban against apples from fireblight-affected areas is under threat, and our ban against the importation of cooked chicken meat has now been modified. It is important that we participate at the international level in forums such as the WTO and the Codex to ensure that health and environmental standards achieve the appropriate precautionary and fair balance. Meanwhile at home, encouraging sustainable agricultural practices will not only allow long-term agricultural ventures to exist, but give us a head start in being able to avoid the pinch of higher standards — fair or otherwise — if and when they are introduced.

Conclusion

The beef hormone dispute might be viewed as a battle of the titans in which Australia was unfortunately entangled. Fortunately for us we were able to comply with the ban to the satisfaction of the Europeans (and get a premium for our meat there as a result) but also we had avenues of trade which were not diminished by the hormone ban.

Yet the implications of the hormone dispute are wider. As the world moves to increasingly liberalised trade with the dismantling of tariffs, we will see increasing reliance upon non-tariff trade barriers by sectors looking for alternative ways to protect their markets. We are also going to see increasing reliance upon similar trade barriers by consumer and environmental groups concerned about the manner in which goods are produced and their health effects.

The SPS Agreement attempts to rationalise protectionist versus precautionary trade barriers on the basis of science. This seems a sensible step as it prevents nations dressing up protectionist causes as environmental and consumer concerns, or overblowing superficial, local or emotive issues without a rational scientific basis. Allowing members to have a reasonable amount of autonomy to choose the level of risk they are willing to bear will probably further the potential for sustainable growth and could slow unsustainable agricultural exports onto world markets.

Future standards are likely to be grounded in environmental concerns relating to the impact of products and may extend to the impacts of their related methods of harvesting, processing and production, for example, standards which impact upon the effect of pesticides on products and possibly farming systems. It is therefore strategically advisable for Australia to recognise this trend and to assist rural development in preparing for it. Australia's natural environment has enabled us in the past to develop an agricultural export market with very little assistance. It is now time to think about how agricultural production can be modified so that again, assisted by our natural environment, we may capitalise on mounting environmental concerns worldwide.

References

Anderson, K., and Blakehurst, R. (Eds) (1992). *The greening of world trade issues.* (Harvester Wheatsheaf: Hemel Hempstead.)

Brenchley, F. (1997). European ban on hormone-treated US beef over-ruled. *The Australian Financial Review* 12 May 1997, p. 14.

Brown, A.J. (1992). Prayers of sense and reason: mining, environmental risk assessment and the politics of objectivity. *Environmental and Planning Law Journal* 4, 387–410.

Burton, I., Kates, R.W., and White, G.E. (1978). *The environment as hazard.* (Oxford University Press: New York.)

Carson, R. (1963). *Silent spring.* (Hamish Hamilton: London.)

Colbourn, T., Dumanoski, D., and Myers, J.P. (1996). *Our stolen future.* (Abacus: London.)

Froman, M.B. (1989). Recent developments in international trade: The United States — European Community hormone treaded beef conflict. *Harvard International Law Journal* 30, 549–556.

Gillette, C.P., and Krier, J.E. (1990). Risk, courts and agencies. *University of Pennsylvania Law Review* 138, 1027–1109.

Grimwade, N. (1996). *International trade policy: a contemporary analysis.* (Routledge: London.)

Halpern, A.R. (1989). The US–EC hormone beef controversy and the Standards Code: implications for the application of health regulations to agricultural trade. *North Carolina Journal of International Law and Commercial Regulation* 14, 135–155.

Handmer, J., Dutton, B., Guerin, B., and Smithson, M. (Eds) (1991). *New perspectives on uncertainty and risk.* (Centre for Resource and Environmental Studies, Australian National University, and Australian Counter Disaster College, Natural Disasters Organization: Canberra and Mt Macedon.)

McDonald, J. (1997). *The WTO's response to the trade-environment debate: a preliminary report card.* 23rd International Trade Law Conference, 29 May 1997, Australian National University, Canberra, ACT.

Report of the Appellate Body on European Communities — Measures Concerning Meat and Meat Products (Hormones) WT/DS26/AB/R; WT/DS48/AB/R.

Report of the Appellate Body on United States — Measures Affecting Imports of Woven Wool Shirts and Blouses from India, WT/DS33/AB/R.

Report of the Appellate Body on United States — Restrictions on Imports of Cotton and Man-made Fibre Underwear, WT/DS24/AB/R.

Report of the Panel on European Communities — European Communities — Measures Concerning Meat and Meat Product (Hormones) — complaint by the United States WT/DS26/R/USA; Complaint by Canada, WT/DS48/R/CAN.

Schroeder, C.H. (1986). Rights against risk. *Columbia Law Review* 86(1), 495–562.

Steinberg, R.H. (1997). Trade-environment negotiations in the EU, NAFTA, and WTO: regional trajectories of rule development. *The American Journal International Law* 91(April), 231–267.

United States Department of Agriculture Food Safety and Inspection Service (1987). *Economic impact of the European Economic Community's ban on anabolic implants.* (USDA: Washington, DC.)

Vogel, D. (1995). *Trading up: consumer and environmental regulation in a global economy.* (Harvard University Press: Cambridge, Massachusetts.)

Wynter, M.C.P. (1996). *Countervailing environmental subsidies in our world trade order.* Ecopolitics X Conference, 26–29 September 1996, Australian National University, Canberra, ACT.

Impediments to the achievement of the commercial and conservation benefits of farm forestry

Alan W. Black

Introduction

Since the late 1980s, various reports and government policy statements have alluded to the substantial economic and environmental benefits that would accrue to Australia by the more widespread adoption of agroforestry or farm forestry (Commonwealth of Australia 1991, 1992, 1995; Farm Forestry Task Force 1995; Greening Australia 1996). This chapter addresses the question of why, despite these potential benefits, there has been a relatively slow rate of adoption of agroforestry or farm forestry in Australia. The usefulness of an explanatory schema devised for more general purposes by Vanclay and Lawrence (1995) will be discussed. It will be argued that the achievement of the potential commercial and conservation benefits of farm forestry depends heavily on the extent to which the force of these impediments can be successfully reduced.

According to the International Council for Research on Agroforestry, '[a]groforestry is a collective name for land-use systems in which woody perennials (trees, shrubs, etc.) are grown in association with herbaceous plants (crops, pastures) and/or livestock in a spatial arrangement, a rotation or both, in which there are ecological and economic interactions between the tree and the non-tree components of the system' (quoted in Prinsley 1991, p. 9). The term 'farm forestry' is sometimes used to refer simply to that component of agroforestry that is concerned with the incorporation of *commercial* tree growing into farming systems; ie where the intention is that a product from the trees (eg sawlogs, pulpwood, eucalyptus oil) will be harvested and sold. However, the term 'farm forestry' is often used in Australia almost synonymously with the term 'agroforestry' as defined above (see, for example, Department of Primary Industries and Energy 1995; Curtis and Race 1996; Race and Curtis 1996; Greening Australia 1996).

The predominant form of *industrial* forestry in Australia is on relatively large tracts of land used more or less exclusively for tree growing, although some industrial plantations occur on land that was formerly used for farming. At the other end of the spectrum, some tree planting occurs predominantly for Landcare or Rivercare purposes, rather than for specifically commercial purposes. As indicated in Figure 9.1, farm forestry fits between these two extremes, involving the incorporation of commercial tree growing into farming systems. It can take a variety of forms: plantations on farms; woodlots; timberbelts; alleys; and wide-spaced tree plantings. The scale of planting can range from relatively small, as in windbreaks and shelterbelts, to moderately large, when extensive sections of a property are devoted to tree growing. The motivation for planting can vary from substantially conservationist to primarily commercial, though the definition used in this chapter requires that commercial considerations, albeit long-term ones, form at least one element of the motivation. Because trees can serve more than one purpose, the category 'farm forestry' partly overlaps that of Landcare plantings on the one hand and industrial plantations on the other (Prosser 1995; Donaldson and Gorrie 1996).

The potential benefits of farm forestry include: adding to the national supply of timber, essential oils and other tree-based products; diversification and increase of farm incomes; synergistic effects on crop, pasture and animal production; amelioration and containment of land degradation; conservation of biodiversity; and reduction of greenhouse gases. Such outcomes are likely to be beneficial not only to individual farmers but also to the wider community; for example, by reducing Australia's reliance on imports and potentially contributing to Australia's exports, thus improving our trade balance.

The Centre for International Economics recently developed an economic model to estimate the costs and benefits likely to accrue to farmers and to the community at large if the optimal share of farmland is devoted to farm forestry. This model indicated that:

> 'The optimal shares of farm land to be diverted to farm forestry differs across zones and with traditional land uses — from 1% for all agricultural land other than broadacre and dairy farmland, to 5% in the pastoral and wheat-sheep zones and for all dairy farmland, and to 10% for broadacre farmland in the high rainfall zone.' (Centre for International Economics, AACM International, and Forestry Technical Services 1996, p. 48)

If these proportions of land were appropriately used for farm forestry, the annual value of farm forestry to Australia once a sustainable harvest is reached would be about $3.1 billion, excluding

Figure 9.1 Farm forestry

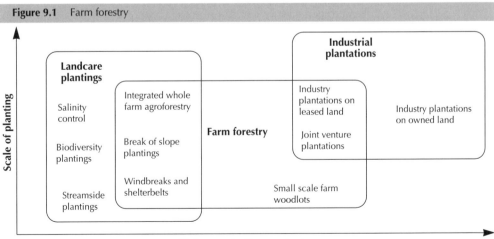

Source: Donaldson and Gorrie (1996)

the value of processed wood products. About 88% of this $3.1 billion would accrue to farmers, but the wider community would also benefit from cleaner water, reduced land degradation and maintenance of biodiversity. In addition, there would be substantial benefits to wood processing industries, the market value of processed wood products from farm forestry increasing to $20 billion a year and employment rising by about 40 000 (Centre for International Economics, AACM International, and Forestry Technical Services 1996, p. 48).

These estimates are based on assumptions of (a) an absence of impediments to the expansion of farm forestry, (b) competitive markets for wood, and (c) constant real prices for wood and for traditional agricultural products. In the quite likely event that there is an increase in the real prices for wood, the value of farm forestry would be even greater than the above estimate. Even if not all the stated conditions are met, farm forestry has the potential to make a substantial contribution to the livelihood of those farmers whose trees have been prudently planted and maintained (Centre for International Economics, AACM International, and Forestry Technical Services 1996, pp. 48–49).

Despite the potential benefits of trees and the emergence of several Commonwealth and State tree planting programs during the 1990s, the uptake of farm forestry in Australia has been much slower than would be required if the commercial and conservation benefits are to be maximised. To estimate the extent of planted trees on Australian farms, the Australian Bureau of Agricultural and Resource Economics (ABARE) conducted a survey of 2000 broadacre and dairy farmers across Australia in 1994. Thirty-five per cent of respondents stated that they had tree belts and corridors ('groups of trees planted in a strip with a maximum width of approximately 30 metres for purposes such as shelter and wildlife conservation'); 14% had tree blocks ('trees grown in high density clumps'); 6% had alley belts ('plantings of at least two strips of trees with grazing or cropping between the strips'); and 6% had widely spaced plantings ('separated and scattered trees on farms, with grazing or cropping carried out between the trees'). The term 'planting' was used to refer to 'a group of at least 20 trees planted together and at the same time in one of the above four planting regimes, with the paddock boundary fence marking the maximum area of any one planting' (Wilson et al. 1995, pp. 9–10).

The ABARE study also revealed that the main functions attributed by farmers to the planting of trees on their farms were to provide shelter and shade, to treat or prevent land degradation, and to conserve native vegetation and wildlife. On less than 2% of farms where trees had been planted in the previous 3-year period was the production of timber or other tree products the main stated function of such trees; in another 4% of cases this was a secondary aim. About a quarter of all broadacre and dairy farmers who had, during the abovementioned period, planted trees received help from some form of assistance scheme for tree planting. Only about 2% participated in a joint venture or share farming agreement for this purpose, reflecting a possible lack of awareness, availability or acceptance of such arrangements (Wilson et al. 1995, pp. 18–20, 23–25).

Over 26 million new seedling trees were planted out on Australian farms during the year ended 31 March 1995 (Australian Bureau of Statistics 1997, p. 69), and almost 19 million in the following 12 months (Australian Bureau of Statistics 1998, pp. 23–24). At the end of March 1996, established plantings of trees on Australian farms occupied a total of approximately 166 000 hectares, which represents 0.036% of the 465.2 million hectares of land used for agricultural and pastoral purposes. This proportion falls a long way short of the optimal proportions quoted above from the Centre for International Economics, AACM International, and Forestry Technical Services (1996, p. 48). Of the 166 000 hectares of established plantings of trees on farms, 51% were in Western Australia and 26%

were in Victoria. Taking Australia as a whole, one-third of the area devoted to such plantings was for timber or pulp production (Australian Bureau of Statistics 1998, p. 24).

Impediments to the adoption of farm forestry in Australia

Four recent studies (AACM International, Centre for International Economics and Forestry Technical Services 1996; Alexandra and Hall 1996; Greening Australia 1996; Curtis and Race 1996) have examined some of the impediments to farm forestry in Australia. Leaving aside technical impediments that have been addressed elsewhere (eg Prinsley 1991), the study by AACM International, Centre for International Economics, and Forestry Technical Services (1996) listed impediments under four headings: lack of a farm forestry culture; economic uncertainties; public policies; and market impediments.

In their analysis, Alexandra and Hall (1996) used three very similar headings: lack of a farm forestry culture; public policy impediments; and market impediments (see also Alexandra and Hall 1998). Their work was in turn used in a publication by Greening Australia (1996). Curtis and Race (1996) identified the following as major socio-economic factors affecting regional farm forestry development in Australia: uncertainty about the economic and environmental viability of farm forestry; uncertainty about the respective roles of growers, industry and government; inadequate understanding of the likely socio-economic impact of farm forestry upon each region; and life-style, life-cycle and socio-demographic factors affecting decision making at farm level.

Prior to these studies, and without focusing specifically on farm forestry, Vanclay and Lawrence (1995) put forward a classification of legitimate reasons for non-adoption of innovative agricultural technologies or integrated management practices, especially those designed to achieve environ-mentally beneficial effects. One might ask: To what extent does Vanclay and Lawrence's schema cover the range of factors influencing the adoption of farm forestry? Each of Vanclay and Lawrence's factors will be considered in turn. Then the question of whether there are other important factors will be considered.

1 Complexity of innovation

It has long been recognised that the rate of adoption of an innovation tends to be negatively related to the perceived complexity of the innovation (Rogers 1983). Successful implementation of farm forestry requires the integration of a diverse range of skills, knowledge and management practices. Some of these requirements, such as silvicultural knowledge and skills, as well as knowledge of timber harvesting, marketing and processing, together with planning horizons ranging from 10 to 30 years, have not hitherto been core competencies in many farming enterprises. Farm forestry is thus an innovation of considerable complexity. This helps to explain its relatively low rate of adoption so far.

2 Divisibility of innovation

Divisibility refers to whether it is possible to adopt a part of an innovation rather than the whole. Divisibility is closely related to what Rogers (1983) termed 'trialability'. Clearly it is possible to grow a few trees without necessarily engaging in commercial farm forestry. However, the success of commercial farm forestry depends in part on the achievement of a scale of operation sufficient to justify the purchase of specialised equipment or, at the very least, sufficient to make it economic to hire contractors for any tasks the farmer cannot perform. Consequently, farm forestry is not as divisible, nor as trialable on a limited basis, as are some other less complex innovations. Another important aspect of trialability is that it takes many years for trees to mature. There is therefore not the same immediacy in judging either the commercial or the conservation benefits of farm forestry. Hence, the speed at which others follow the lead of early adopters is likely to be correspondingly slower than would be the case for innovations whose commercial or conservation benefits are more quickly manifest.

3 Congruence or incompatibility with farm and personal objectives

Under this heading, Vanclay and Lawrence referred to the degree of compatibility both with existing farm enterprises and with other personal or family goals. One possible reason for adopting farm forestry is because of its synergistic effects on the production of animals and other plants. Thus, the Centre for International Economics, AACM International, and Forestry Technical Services (1996, p. 48) have calculated that improved yields resulting from shelter belts in the sheep–wheat zone are sufficient to cover the costs of establishing and maintaining such shelterbelts, even if no wood is sold. Nevertheless, the demands of farm forestry in terms of land, labour and capital may not always be compatible with other farming activities or with other personal or family priorities. The availability of family labour for operations such as planting, pruning, spraying and thinning may facilitate the adoption of farm forestry by minimising expenditures on these operations. Conversely, the absence of such labour, coupled with low incomes from existing enterprises whether on or off the farm, is likely to impede the adoption of farm forestry. So, too, may other family priorities such as the education of children. On the other hand, off-farm income may provide resources that enable the adoption of farm forestry even though the economic returns will be delayed. Thus family dynamics, including family size, life-stage, gender, likely inheritance patterns and extent of off-farm employment, may significantly influence the adoption process (Curtis and Race 1996).

4 Effect on farmers' flexibility

Because trees take much longer to reach maturity than most other farm commodities, farm forestry imposes some limitations on farmers' capacity to respond to changing market or climatic conditions. It is obviously much more difficult to change the species of trees one is growing than to switch from producing one variety of grain to another. Likewise, in the event of a downturn in the market for timber, it would generally be more difficult to redeploy the land used for timber production than to switch from one broadacre enterprise to another. Some farmers shy away from farm forestry because of this perceived loss of flexibility. The effect of this deterrent may be reduced through a wider awareness of the synergistic benefits of farm forestry, as well as by more reliable forecasts of future supply of, and demand for, particular species of timber.

5 Likely economic benefits

Under this heading, Vanclay and Lawrence note that although economic considerations are not the only factor in adoption decisions, it is reasonable to expect that, everything else being equal, the more economically beneficial an innovation, the greater the rate of adoption; and that practices with more immediate benefits tend to take precedence over those with only long-term benefits. Rogers (1983, pp. 15, 213) used a broader concept, relative advantage, which he defines as 'the degree to which an innovation is perceived as better than the idea it supersedes'. He observes that although economic benefit is the most common measure of relative advantage, other considerations such as social prestige, convenience or satisfaction may also be important. Clearly, perceived economic benefits, both short-term and long-term, are major determinants, though not necessarily the only determinants, of the likelihood that farmers will adopt farm forestry.

6 Implementation cost: capital outlay

Wilson *et al.* (1995, pp. 21–22) found that establishment costs for farm forestry vary between about $700 and $1800 per hectare, depending on the planting regime involved; labour (calculated at $12.50 per hour) and fencing are the main components of these costs. Eckersley, Ellis and George (1993) put the cost of establishing tree blocks as being typically between $1200 and $1500 per hectare although they noted that establishment costs might be reduced to $600 per hectare if existing farm labour and equipment can be used. These establishment costs are much higher than those for perennial pasture, which Francis and Marcar (1993) calculated as ranging from $200 to $250 per hectare. High establishment costs, coupled with a long wait for a return on the investment, pose a major problem for many farmers, especially if they must borrow to finance the venture.

7 Implementation cost: intellectual outlay

Implementation cost in terms of intellectual outlay refers primarily to the amount of new learning required by the individual farmer. This is closely related to the complexity of the innovation. By current standards, many farmers have had a rather limited formal education and many are now aged between 50 and 70 years. Without implying that persons in these categories are necessarily unwilling or unable to learn new competencies, there are obviously limits to the extent to which they are ready to acquire the wide range of skills and knowledge required for successful farm forestry. This situation poses a particular challenge for persons who are responsible for designing and implementing farm forestry extension and training programs.

8 Risk and uncertainty

There are various risks and uncertainties associated with farm forestry. Some of these relate to contingencies of production such as weather and the prevalence of tree pests or diseases, some relate to markets for logs and for other tree products, and some relate to public policy in areas such as taxation, local government rates, land use planning controls, tree tenure, harvesting rights and export controls. As these issues have been examined in depth elsewhere (eg AACM International, Centre for International Economics, and Forestry Technical Services 1996; Alexandra and Hall 1996; Greening Australia 1996), I shall not explore them in detail here.

In so far as the risks and uncertainties mentioned above continue, they are impediments to the adoption of farm forestry. Fortunately, some of these impediments have recently been removed or reduced by the implementation of the Wood and Paper Industry Strategy of the Commonwealth of Australia (1995). In particular, the Federal Government has taken steps to remove anomalies and uncertainties about various taxation matters. At its July 1996 meeting, the intergovernmental Ministerial Council on Forestry, Fisheries and Aquaculture affirmed the intention of the various States and Territories to enact legislation that would remove uncertainty over the right to harvest wood grown for commercial purposes and that would provide a sound legal basis for separating the ownership of trees from the ownership of the land on which they grow (Anderson 1996). Such legislation would eliminate or reduce some of the risks and uncertainties that have been associated with farm forestry hitherto.

9 Conflicting information

In Western Australia, the Farm Forestry Task Force (1995, p. 3) reported that 'It is hard to access good information on a whole farm forestry package from independent sources. There is some evidence of farmers receiving conflicting advice from different sources'. Likewise, Alexandra and Hall (1996, p. 10) noted that the possibility of receiving conflicting information on farm forestry is accentuated because there is inadequate co-ordination between agriculture and forestry departments or between different divisions of amalgamated departments. These differences of orientation are often reinforced by different emphases in the tertiary curricula for foresters and agriculturalists respectively.

The receipt of conflicting information is clearly an impediment to the adoption of farm forestry. However, it does not follow from this that the receipt of consistent information necessarily increases the rate of adoption. Consistent information about the potential problems of farm forestry might act as a deterrent to adoption, especially if it is not counterbalanced by reliable information about farm forestry's potential benefits.

10 Perceptions of the environment

Farm forestry has been promoted not only for its commercial benefits but also for its environmental benefits, especially for its potential to prevent or alleviate problems of dryland salinity. While it is reasonable to expect that the likelihood of the adoption of environmentally beneficial innovations increases when farmers consider that their own farms are at risk from environmental degradation, studies cited by Vanclay and Lawrence (1995, pp. 82–86) indicated that farmers are often unaware of the early warning signs of land degradation and they usually understate and misperceive the extent to which their own properties are affected by such degradation. Media images of land degradation, whether on television, in newspapers, in conservation organisation publications and in some agricultural extension literature, tend to focus on the most severe and visually obvious forms of land degradation. As many farmers may not currently be experiencing land degradation of such severity, the media images may unintentionally lull them into a false sense of security rather than prompting them to alleviate existing degradation and prevent further degradation.

Furthermore, in the case of dryland salinity, those farms on which soil salting is most evident, such as those in low lying areas, may not be solely or even primarily responsible for the problem. To deal effectively with the problem, trees may need to be planted not only on the farms where the problem is most evident but also on farms where the problem is not so evident (Vanclay and Lawrence 1995, pp. 90–91). This need for concerted action poses an additional challenge for those responsible for farm forestry extension programs.

11 Physical infrastructure

Physical infrastructure requirements for farm forestry include not only supplies of various inputs but also access to appropriate markets for logs and for other tree products. Because, all other factors being equal, the price achievable at the mill gate is determined by supply and demand, differences in transport costs directly affect the net return to growers. Consequently the availability and efficiency of road, rail and port services, as well as of processing facilities, are important aspects of the physical infrastructure. These facilities in turn require adequate power supply, water and waste disposal services.

Vanclay and Lawrence also regarded the availability of contractors or a skilled labour force, of consultants and extension personnel, as well as of other sources of information, as part of the physical infrastructure. Although these things have a physical element to them, they could perhaps better be considered as part of the required social infrastructure.

12 Social infrastructure

Under this heading, Vanclay and Lawrence referred primarily to the existence of a sufficient degree of interest in an innovation to promote widespread adoption. This is clearly important for the achievement of economies of scale in marketing or processing in any particular region, as well as to maximise conservation benefits. Organisations such as Landcare groups, farm forestry networks, grower co-operatives, marketing agencies and the other forms of social infrastructure mentioned in the previous paragraph are also likely to be important for the adoption of farm forestry.

13 Farming subcultures and styles of farming

Although most farmers prize their capacity to be independent, they are also influenced by the farming subculture in which they are located. This subculture consists of the prevailing beliefs, attitudes and way of life within the farming community. Some older farmers who were brought up at a time when tracts of land were being cleared for agricultural purposes are averse to the suggestion that some land should now be revegetated with trees. Many farmers are strongly opposed to any moves to turn whole farms in their vicinity into forestry plantations, as they see this as possibly leading to further rural depopulation, with a consequent decline in local social amenities (Rose 1996). As those purchasing land for industrial forestry are often able to pay higher prices than are nearby farmers, the latter find it difficult to expand their operations. Opposition to industrial forestry on these and other grounds (such as the view that it is wrong to take land out of food production when there is a world food shortage) can colour farmers' attitudes toward the planting of trees on their own farms.

Although some farming subcultures and farming styles militate against the adoption of farm forestry, the potential is also there to work with opinion leaders, using group processes to engender more widespread recognition of the benefits of farm forestry and to encourage more widespread practice of it.

Concluding comments

With some modifications and elaboration, Vanclay and Lawrence's schema provides a useful framework to analyse many of the impediments to the adoption of farm forestry. However, some specific factors that apply to farm forestry do not necessarily have exact parallels elsewhere. For example, farm forestry involves a much longer time from planting to harvest than does the production of most other commodities. This, together with the relatively high capital outlay required, poses particular difficulties for many farmers. While there are various solutions to these problems, such as joint-venture, leasing and share-farming arrangements, each of these needs to be carefully considered before being entered into. Many farmers are very cautious about 'getting into bed' with big business. Furthermore, farm forestry is the only farm-based activity that has to face large-scale direct (and, in at least some cases, almost monopolistic) competition from government agencies, namely departments involved in the ownership of industrial forest plantations. Although some types of farm forestry could develop niche markets, there are obviously limits to the extent to which this is possible.

In principle, farm forestry has the capacity to make a substantial contribution to the achievement of various conservation objectives. One such objective relates to the supply of wood. World demand for structural timber is increasing faster than supply. Various Asian countries have been rapidly depleting their domestic sources of wood. By the year 2000, Pacific Rim countries will be facing an estimated annual timber shortfall of 325 million cubic metres, which is equal to one-fifth of the current world trade in industrial wood (Ryan 1994, quoted in Curtis and Race 1996, p. 5). To help achieve a sustainable supply of wood and wood products for both domestic and export purposes, the timber industry recently set itself the goal of trebling Australia's forest plantation estate by the year 2020. That goal was endorsed by the Australian government in its Wood and Paper Industry Strategy (Commonwealth of Australia 1995). It was envisaged that a substantial part of the increase would come from new farm forestry plantations. Farm forestry also has the potential to contribute to other conservation objectives, such as reducing the rate of salinisation by lowering the watertable; reducing greenhouse gases through the sequestration of carbon dioxide; and conserving biodiversity, especially of native flora and fauna.

The achievement of the potential commercial and conservation benefits of farm forestry depends heavily on the extent to which the force of the impediments discussed in this chapter can be successfully reduced. In addition to making available reliable technical information on farm forestry, there is a need to address the public policy, marketing, infrastructural and cultural impediments outlined above. Fortunately, action has already been initiated on some of these fronts (Donaldson and Gorrie 1996; Ministerial Council on Forestry, Fisheries and Aquaculture, Standing Committee on Forestry, Plantations Australia, Australian Forest Growers, National Association of Forest Industries 1997; Alexandra and Hall 1998). Others will require initiatives not only by Federal, State and local governments but also by groups of farmers and by other investors.

References

AACM International, Centre for International Economics, and Forestry Technical Services (1996). *Commercial farm forestry in Australia: development of a strategy framework — a synopsis.* (Rural Industries Research and Development Corporation, Land and Water Resources Research and Development Corporation, Forest and Wood Products Research and Development Corporation: Canberra.)

Alexandra, J., and Hall, M. (1996). *Creating a viable farm forestry industry in Australia — what will it take?* Discussion paper (mimeo).

Alexandra, J., and Hall, M. (1998). *Creating a viable farm forestry industry in Australia.* (Rural Industries Research and Development Corporation, Land and Water Resources Research and Development Corporation, Forest and Wood Products Research and Development Corporation: Canberra.)

Anderson, J. (1996). *Australia to treble its forest plantations by 2020.* Media release DPIE96/59A of 29 July 1996, issued by the Minister for Primary Industries and Energy.

Australian Bureau of Statistics. (1997). *Agriculture Australia 1994–95.* (Australian Bureau of Statistics Catalogue No. 7113.0: Canberra.)

Australian Bureau of Statistics. (1998). *Agriculture Australia 1995–96.* (Australian Bureau of Statistics Catalogue No. 7113.0: Canberra.)

Centre for International Economics, AACM International, and Forestry Technical Services. (1996). *Contribution of farm forestry to Australia: a quantitative assessment.* (Rural Industries Research and Development Corporation, Land and Water Resources Research and Development Corporation, Forest and Wood Products Research and Development Corporation: Canberra.)

Commonwealth of Australia. (1991). *Integrating forestry and farming: commercial wood production on cleared agricultural land.* Report of the National Plantations Advisory Committee: Canberra.

Commonwealth of Australia. (1992). *National forest policy statement: a new focus for Australia's forests.* Canberra.

Commonwealth of Australia. (1995). *Wood and paper industry strategy.* (Forests Taskforce, Department of the Prime Minister and Cabinet: Canberra.)

Curtis, A., and Race, D. (1995). *Review of Department of Primary Industries and Energy National Farm Forestry Program,* Vols 1–2.

(The Johnstone Centre of Parks, Recreation and Heritage, Charles Sturt University: Albury, NSW.)

Curtis, A., and Race, D. (1996). *A review of socio-economic factors affecting regional farm forestry in Australia.* (The Johnstone Centre of Parks, Recreation and Heritage, Charles Sturt University, Report No. 69: Albury, NSW.)

Department of Primary Industries and Energy. (1995). *Plantations and farm forestry.* (Forests Branch, Department of Primary Industries and Energy: Canberra.)

Donaldson , J.D., and Gorrie, G.C. (1996). *Farm forest policy.* Paper presented at Australian Forest Growers Conference, Mount Gambier (mimeo).

Eckersley, P., Ellis, G. and George, R. (1993). *Bluegums: a real option.* (Department of Agriculture: Perth.)

Farm Forestry Task Force (1995). *Farm forestry in Western Australia.* (Farm Forestry Task Force: Perth.)

Francis, J. and Marcar, N. (1993). *Dryland salinity: productive use of salt affected land.* (Department of Conservation and Land Management, NSW, and CSIRO Division of Forestry: Canberra.)

Greening Australia 1996. *Farm forestry in Australia: integrating commercial and conservation benefits.* (Greening Australia: Canberra.)

Ministerial Council on Forestry, Fisheries and Aquaculture, Standing Committee on Forestry, Plantations Australia, Australian Forest Growers, National Association of Forest Industries. (1997). *Plantations for Australia: the 2020 vision.* (Plantation 2020 Vision Implementation Committee: Canberra.)

Prinsley, R. (1991). *Australian agroforestry: setting the scene for future research.* (Rural Industries Research and Development Corporation: Canberra.)

Prosser, M. (1995). *An industry perspective on farm forestry.* Paper developed from a presentation to Murray Farm Forestry Meeting, December (mimeo).

Race, D., and Curtis, A. (1996). Farm forestry: how things stand. *Australian Journal of Soil and Water Conservation* **9**(2), 29–35.

Rogers, E.M. (1983). *Diffusion of innovations,* 3rd edn. (Free Press: New York.)

Roling, N. (1988). *Extension science: Information systems in agricultural development.* (Cambridge University Press: Cambridge.)

Rose, B. (Ed) (1996). *Bridgetown-Greenbushes and Boyup Rural Survey, September 1996. Results and Conclusions.* Lithographed.

Vanclay, F., and Lawrence, G. (1995). *The environmental imperative: eco-social concerns for Australian agriculture.* (Central Queensland University Press: Rockhampton.)

Wilson, S.M., Whitham, J.A.H., Bhati, U.N., Horvath, D., and Tran, Y.D. (1995). *Survey of trees on farms 1993–94.* (Australian Bureau of Agricultural and Resource Economics Research Report 95.7: Canberra.)

Toward regional strategies for rural sustainability: a farmer's view

Peter Milliken

Introduction

My family house stands some metres from the riverbank near Hay on the Murrumbidgee River in southern New South Wales. The 1956 flood saw an intrepid 3-year-old sailing a washtub in our front yard as the water lapped the floorboards. My childhood leisure was largely spent on the water in a canoe fishing for cod, yellowbelly, bream, catfish and later redfin. I learnt the patterns of the river, the cyclical changes and realised its fundamental importance. I also watched the changes.

In later years I studied at Sydney University and taught at Wagga Wagga High School and watched other kids who seemed to have no sense of origin or attachment to a landscape, and wondered.

Later still, I realised that there was no one to speak for the Murrumbidgee Catchment with any sense of ownership or continuity. Government agency personnel had various unrelated, and sometimes conflicting, roles in natural resource management. These personnel have come and gone with career moves while the community felt powerless to respond.

The advent of Landcare (Farley and Toyne 1989) and Total Catchment Management (eg Anon 1989) in the Murrumbidgee River catchment changed this and allowed the development of community ownership. The process of linking community and government actions in natural resource management began. My involvement as a grazier and irrigator put me in the melting pot of change.

Murrumbidgee Catchment Management Committee

The formation of the Murrumbidgee Catchment Management Committee (MCMC) in 1990 brought interested community people like myself together with personnel from government agencies, local government and environmental interest groups. Our meetings have traversed the 84 000 square kilometres of the catchment from Cooma to Balranald including the Australian Capital Territory (ACT). The MCMC developed a Natural Resource Management Strategy in 1993 (Murrumbidgee Catchment Management Committee 1994) and in some ways, most of the 520 000 people living in the catchment have been affected by our actions and are becoming increasingly aware (albeit slowly) of the terms 'Total Catchment Management' and 'Integrated Resource Management'.

The experience, current actions and broad representative base of the MCMC positioned us well to volunteer as a pilot catchment for the development of a Regional Strategy under the Natural Heritage Trust processes.

A change of process under the Natural Heritage Trust

Previously the catchment management committee operated with funding from a variety of State and Federal sources. In 1997, the Commonwealth established the Natural Heritage Trust (NHT) to direct the reinvestment of funds into actions targeting sustainable resource management. The NHT consists of a package which includes Commonwealth, State and Territory money organised through partnership arrangements. The Murrumbidgee Catchment is one of three catchments throughout the State to pilot the Regional Strategy component of the NHT.

In the past, one of the roles of the MCMC was passive assessment and prioritisation of applications for the National Landcare Program (NLP) against the issues outlined in our Natural Resource Management Strategy. The process developed under the NHT is much more proactive. Our Regional Strategy (Murrumbidgee Catchment Management Committee 1998) examines a wide range of investment sources and analyses our natural resource issues and current actions. We then build an investment portfolio for the Murrumbidgee that is both balanced and leverages external investment for the most cost-effective outcomes in the catchment. We need actions that are self-sustaining to build our investment in natural resource management. We intend to clearly indicate areas for investment and actively commission works to ensure a balanced investment in our catchment. To operate in this manner the MCMC, I believe, will eventually operate under a trust or incorporated structure. As the level of direction in finances increases, so the need for account-ability increases. Monitoring of outcomes is needed to ensure completion is both timely and cost and ecologically effective.

The Murrumbidgee Natural Heritage Trust Committee (MNHTC) was formed from members of the MCMC, and began the task of creating a Regional Strategy for the Murrumbidgee Catch-ment in early 1997.

As always, a large committee is constrained by many factors. So it came down to four volunteers, supported by a small team funded by the MCMC, to create a draft document. The committee represented a mix of expertise and included representatives of the three major resource management agencies — the Department of Land and Water Conservation, the National Parks and Wildlife Service, the Parks and Conservation agency of the Australian Capital Territory — and myself as a farmer.

The final version of the draft Regional Strategy (see below) was completed in mid-1997 and made available for public comment at a number of public forums throughout the catchment. This period of consultation has allowed the broad community to debate the vision and the detailed investment plans outlined in the document.

The short time frame for development of the strategy is evident, and involved a heavy workload and dynamic drafts. It created some resistance to change from committee members and a questioning of roles and responsibilities — a healthy part of the debate.

The MNHTC will play a key role in ensuring competitive tendering delivers cost effective outcomes for natural resource investment. Below, I have outlined the principles we used in developing the Regional Strategy, the objectives of the strategy, and how we will go about making it work.

The regional strategy

Vision and objectives

There are a number of international, national and State responsibilities which the Regional Strategy recognised and sought to achieve. These included:

- *Agenda 21* (an agreement to implement environmental objectives at the local government level, a clause of the Rio Convention);

- *Intergovernmental Agreement on the Environment* (IGAE) (sets out the responsibilities of Federal, State, Territory and local governments with respect to environmental policy and management); and,

- *Council of Australian Governments (COAG) reform agenda* (a set of reforms covering water pricing, entitlements, institutional reform, and public consultation and research).

In addition, the National Strategy for Ecologically Sustainable Development (Commonwealth of Australia 1992) aims to meet the needs of Australians today, while conserving our ecosystems and the ecological processes on which life depends, for the benefit of future generations. It means developing ways of using those natural resources which form the basis of our economy in a way which maintains and, where possible, improves their range, variety and quality. At the same time we need to utilise those resources available to develop industries and employment.

The core objectives of the National Strategy for Ecologically Sustainable Development are:

- *to enhance individual and community welfare by following a path of economic development that safeguards the welfare of future generations;*
- *to provide for equity between generations; and,*
- *to protect biological diversity and maintain ecological processes and life support systems.*

We used these international and national agendas and strategies to guide us in framing the following vision for our catchment:

A productive Murrumbidgee Catchment with healthy ecological processes and enhanced biodiversity.

We used the general objectives of the National Heritage Trust (Commonwealth of Australia 1998) as a guide to setting the actions and outcomes for our Regional Strategy (Table 10.1).

The Regional Strategy thus aims to create an investment portfolio for the Murrumbidgee Catchment based on social, economic, and environmental criteria to achieve ecologically sustainable management. This means that investors and stakeholders will know where and why investment is being encouraged, and what the returns on investment are likely to be. The Regional Strategy will encourage partnerships among environment and natural resource management stakeholders, and encourage the growth of viable businesses based on ecologically sustainable development (ESD).

Achieving the objectives

There are a number of ways in which the strategy will work towards meeting these objectives. These activities require the collaboration of various groups, which are committed to the Regional Strategy.

Table 10.1 Objectives of Natural Heritage Trust and corresponding regional actions and outcomes.

Objectives	Regional actions and outcomes
To facilitate the development of strategic approaches to regional issues; encourage investment for achievement of ecologically sustainable development (ESD)	The development of the Murrumbidgee Catchment Action Plan and its implementation through community, agencies and private sector
To stimulate additional public and private investment in overcoming barriers to achieving ESD	Promoting the goals of ESD, and encouraging investment in integrated projects which address social, economic and environmental issues through specific business plans (eg ecotourism business plan)
To enable natural resource, biodiversity conservation and environment protection priorities to be addressed effectively and efficiently	The Regional Strategy identifies existing impediments and alternative approaches to management and the programs to enhance, such as agency/MCMC interaction
To encourage partnerships among environment and natural resource management stakeholders	The region is developing a marketing and communication strategy, encouraging partnerships among the community, private sector and government

The Regional Strategy will:

- adapt the Natural Resource Management Strategy for the Murrumbidgee Catchment (MCMC 1994) into **an investment strategy** for ESD;
- *develop* priorities for natural resource action in the Murrumbidgee Catchment through the development of the **Murrumbidgee Catchment Action Plan** (MCAP);
- *encourage* the stewardship of all the resources of the catchment, towards everyone taking responsibility for a healthy environment and a sustainable future;
- ensure that the management of natural resources and the environment in the region is optimal through local planning, improved land use, and by adopting **best management practices** in conservation, environment protection and in primary production;
- *investigate and promote* **low cost alternatives** in natural resource management;
- seek to strengthen rural communities and *provide networks* for landholders and agencies;
- serve as the **prospectus for investors** seeking to invest in sustainable environmental and natural resource management in the Murrumbidgee region, and provide a framework for other groups and agencies to link programs to employment and so on.

A community group can use the Regional Strategy to understand the regional and State-wide context of issues affecting their catchment. This will assist in targeting their efforts at attracting investment which will address the priority issues for the catchment.

Local government and State Government agencies will be able to better direct their resources in co-ordination with each other and the community.

The Regional Strategy will ensure links are established with existing programs, for example, for ecotourism: 'Through responsible planning and management practices and product development, regional Australia can provide tourism opportunities that will deliver long-term local employment and business growth' (Office of National Tourism 1996).

Investment strategy

The Murrumbidgee Regional Investment Strategy identifies where investment will be directed. Investment will be in projects addressing the priority areas identified in the strategy. The investment will be in projects dealing with the causes and not the symptoms. The type of projects for investment will be those that clearly demonstrate the following categories of activity and that adequate assessment and planning have been carried out to enable implementation: assessment; education and awareness; planning; implementation; and monitoring.

The funds available under the current arrangements of the NHT form the venture capital or bridging finance to enable a viable industry based on ESD and natural resource management. Future years of the Murrumbidgee Regional Strategy will focus on such activities, allowing the MCAP to develop links with ongoing projects and new initiatives.

There will be an emphasis on implementation of those plans which have been completed and assessed as being high priority. New planning projects will be in accordance to the MCAP sequence.

Agency, local government and private sector investment will be linked to the NHT investment, to ensure that there is a concerted effort in mutually agreed directions.

Agencies will be reviewed to assess the outcomes from their natural resource programs, and reporting arrangements for projects will ensure that they are tracked and outcomes are achieved.

The Regional Strategy also recommends specific projects to address gaps which are critical to the sustainability of the catchment. In future years a process will be established which enables early identification of gaps and commissioning of projects to address these. However despite limited time this year a gap has been identified in the area of projects addressing biodiversity management. A proposal to promote ecotourism focused on biodiversity in the catchment has been developed for inclusion in the 1997–98 funding bids to the NHT. The proposal includes education and training, marketing and involvement of the private sector in establishing viable ecotourism-based projects for the whole catchment.

The first step being undertaken under the MCAP is to identify the agency programs in natural resource management and match their outcomes to the Regional Strategy. Conservation projects are placed on high priority, provided there is a clear process of ensuring the long-term self-sufficiency of the projects.

Murrumbidgee Catchment Action Plan (MCAP)

The MCMC has developed a draft Murrumbidgee Catchment Action Plan (MCMC 1998) which will be a key element in meeting the objectives of the Regional Strategy. The MCAP will provide direction to agencies to operate with a whole of government approach, and to community and other resource managers in *resource access, operation, improvement* and *management*. It will set the boundaries and conditions in consideration of upstream influences and downstream impacts.

The MCAP translates the MCMC's Natural Resources Management Strategy, the Regional Strategy and other community and State agency strategies into a plan of action to achieve outcomes. The MCAP will provide a regional framework from which on-ground works and local action plans can be developed.

The draft MCAP was released for public consultation and debate in early 1998. Ongoing consultation with the MCMC during the further development of the Regional Strategy and the MCAP will ensure that consistent objectives and complementary processes are established, to provide for the long-term sustainability of the catchment.

Separation of power and responsibilities

A key feature of the process is the separation of roles and responsibilities of those developing a Regional Strategy from those assessing the annual bid for investment through project applications. With this in mind, a Regional Assessment Panel will be formed, which will be entirely independent of the Murrumbidgee National Heritage Trust Committee.

Identifying and prioritising issues for action

Because there is such variation in topography, climate, vegetation and land use in the Murrumbidgee Catchment, the natural resource and environment issues also vary in the emphasis and urgency with which they need attention. However, despite the differences there are many issues which are common to the whole catchment. These include the management issues of native vegetation, management of weeds and pest animals, prevention and control of erosion, loss of soil nutrients to waterways, quality of the water and the amount of water available. In addition, the social and economic aspects of resource management such as rural adjustment and property management planning are identified as issues which need to be addressed in parallel with environmental implications.

One of the key principles of ESD is the linking of social, economic and environmental objectives both in policy and practice. The sustainability of a resource base, such as a catchment, is determined by the natural resource characteristics, their interaction, the rate of change, and the environmental, social and economic characteristics of the catchment. Integrated resource management plays a key role in achieving long-term sustainability in the region. By adopting an integrated approach to resource management, the interrelationships between resources are explicitly recognised. This can apply at the catchment scale through to individual property planning which considers the environmental, social and economic factors when addressing resource management.

Previous reviews carried out as part of communication and planning activities within the catchment provided a number of priority areas in the environmental, economic and social spheres (eg Murrumbidgee Catchment Management Committee 1994). These are grouped in the Regional Strategy, and the main threats and opportunities for managing these sustainably for the catchment are presented.

The first step undertaken under the MCAP was to identify the agency programs in natural resource management and match their outcomes to the Regional Strategy. Conservation projects are placed on high priority, provided there is a clear process of ensuring their long-term self-sufficiency.

To help in prioritising bids for funding under the NHT umbrella in 1997, we first compiled a list of resource management issues based on a number of meetings with stakeholders throughout the catchment. Issues were tabulated according to priority, and a table which identified each of the issues (such as dryland salinity, water quality, and soil erosion) was then developed and was used as a guide for assessment of this year's current funding applications to link projects and ensure rigorous analysis for integrated resource management. We focused on the following when assessing projects.

1   *Assessment of priority areas:*   An assessment was made on the *level of work currently carried out* in each category area: assessment, education, planning, implementation and monitoring. This is presented in the table and should be read as follows:

   • key problems associated with the particular resource (eg the primary concern in relation to surface water quality is salinity);

- level of work done towards addressing problem (eg has there been **assessment** of the problem, are people **educated** and **informed** about the problem, is there **strategic planning** for this problem, have any programs been **implemented**, is there **monitoring** to assess effectiveness of actions?); and

- the area of the catchment affected (eg upper-, mid-, lower-catchment);

2 *Initial project assessment:* From this information an initial assessment of current project applications has been made, identifying:

- key areas addressed (eg assessment, education, planning, implementation or monitoring); and

- catchment area covered.

3 *Recommended action:* This section of the table provides commentary on the types of projects which would best meet the regional priorities identified in the Regional Strategy. These recommendations are based on priority issues, geography, current resourcing and past investigations into causes and effects.

The current Regional Strategy utilised the applications lodged under the NHT Program, Natural Resource and Environment Program NSW, to indicate likely actions for next year.

Monitoring and evaluation

The Regional Strategy will be monitored by the Murrumbidgee Natural Heritage Trust Committee. This will ensure that the objectives of the region and government are met. Regular updates of the progress will be presented by key project managers at MNHTC meetings every 2 to 3 months. An independent review of the Regional Strategy will be undertaken every 3 years to report on the returns on previous investment. Based on this review, a portfolio for continuing and new investment by the community, private sector and governments will be prepared. In addition, an annual report will be compiled on outcomes and investments.

Geographic Information Systems (GIS) is a management tool to be used for modeling issues, and the levels of resourcing associated with these. The use of GIS will form an integral part of the monitoring and evaluation of the Regional Strategy and later the MCAP which will contribute to follow-up and reporting.

Conclusion

Why would a farmer want to be involved? I've outlined ownership of the catchment; there are also responsibilities to the future of my four children. However, it also keeps me abreast of change, allowing assessment of future directions — in effect, my adult education.

The posturing and negativity of some of our professional politicians in their two-party political system concerns me. The quality of political debate and the lack of 'independent statesmen' (read 'statesperson') concerns me. I enjoy the apolitical and collaborative approach to determining our future that is evident in Total Catchment Management debate.

I believe farmers have a very valid role to play in strategy development. Their hip pocket nerves and property values are directly affected by their decisions and experience in natural resource management. Farming is a very complex business which involves regular and active decision making, based on social, economic and environmental factors. An experienced farmer can make these decisions intuitively when well supported by information transfer. These decisions drive concrete actions that Mother Nature plays with, to produce variable outcomes. Analysis of the outcomes helps the farmer to refine future decision making.

With this background the NHT process was immediately clear and logical to me. The need to analyse; to concentrate funds on effective actions; to invest in self-sustaining actions; to achieve cost-effective outcomes; to monitor outcomes and to analyse and plan again are familiar steps for me. They represent a very powerful method of achieving our goals.

The National Landcare Program was excellent for its time, but it represented a disbursement of charity dollars to the environment. The Natural Heritage Trust is more soundly based on the interaction of society, economics and environment. It is the beginning of real investment in our future and a mechanism of achieving on-ground outcomes that some other nations can only dream about.

As a farmer I have practised a very hands-on form of production and sale of tangible items such as a kilogram of beef, wool, lamb or wheat. The enormous productive output of the Murrumbidgee Catchment deserves special attention and the drive for ecological sustainability will produce new management systems for traditional agricultural production. I firmly believe it will also produce new industries based around natural resource interpretation and management. I clearly remember the Principal from my childrens' public school saying that '90% of the jobs these children will do have not even been invented yet'. Australia is at the cutting edge of the development of these philosophies and technologies. Our Landcare movement and land and water management planning processes are unparalleled for cost-effectiveness of on-ground actions harnessing community ownership.

In our catchment, we already have groups of farmers like the Rural Awareness Property Tours (RAPT) at Harden beginning to create industries;[1] and people like Sharon Stacy at Tumut developing national teaching protocols for natural resource interpretation.

This catchment is positioned to grasp the opportunities for beneficial change. We intend to achieve a whole-of-government approach focused by a catchment community who want to build a future for their children. This is a powerful force to achieve our common goal. Historically we have not always worked to a common purpose.

If we involve the key players we can achieve our vision of 'a productive Murrumbidgee Catchment with healthy ecological processes and enhanced biodiversity'.

References

Anon (1989). *Catchment Management Act*. (New South Wales Government.)

Commonwealth of Australia (1992). *National strategy for ecologically sustainable development*. (Australian Government Publishing Service: Canberra.)

Commonwealth of Australia (1998). *National Heritage Trust guide to new applications 1998–1999*. (Australian Government Publishing Service: Canberra.)

Farley, R., and Toyne, P. (1989). A national land management program. *Australian Journal of Soil and Water Conservation* **11**, 6–9.

Murrumbidgee Catchment Management Committee (1994). *Natural resource management strategy for the Murrumbidgee Catchment*. (Murrumbidgee Catchment Management Committee: Wagga Wagga.)

Murrumbidgee Catchment Management Committee (1998). *Draft Murrumbidgee Catchment action plan for integrated natural resource management*. (Murrumbidgee Catchment Management Committee: Wagga Wagga.)

Office of National Tourism (1996). *Taking tourism to regional Australia. Office of National Tourism — Projects in profile 1996*. (Office of National Tourism: Canberra.)

1   RAPT organises bus tours aimed at Landcare and Special Interest Groups who want to see how farmers are solving land degradation in a practical way, with a positive result. This enables visiting groups to gain hands-on experience with problems that could occur in their own area. For further information contact Libby Elliot, RAPT Co-ordinator, phone/fax: 06 227 4220.

Saline politics: an inland city case study

Petrina Quinn and Mark Conyers

Introduction

As Australians attempt to find solutions to local and regional environmental problems, we are faced with determining the most appropriate model for community action that will ensure sustainable development. What combination of State-controlled, State-initiated or participatory 'bottom-up' approaches is most effective in rehabilitation projects is a key question for all of us (Curtis 1997). We are increasingly seeing in Australia an encouragement of, and recognition that, local participatory approaches to environmental problems are desirable for both economic and political reasons. The participatory approach is cost-effective from the government's perspective because non-government resources are used. Such approaches are more democratic since a range of people are involved in the decision making and because individuals who would not usually have the opportunity or be in a position to contribute in a political sense are involved (Martin and Woodhill 1995).

In rural regions, environmental problems which occur over large temporal and spatial scales may not be amenable to rehabilitation efforts solely by community groups (Martin and Woodhill 1995) and must be tackled through partnerships between the community and government with appropriate cost-sharing arrangements (Curtis 1997). Such arrangements are a shift from the paradigm that the State will provide and recognise the social context of rehabilitation efforts.

While such shared approaches are recognised as being the most appropriate model for action, recent deregulation policies of the Federal Government and rationalisations by State governments has seen the withdrawal of a range of services from rural Australia. Where once government officers lived and worked in rural communities, increasingly officers have not been replaced once they leave, or jobs are being contracted out on flexible, short-contract arrangements. In such cases government commitment inevitably decreases. A consequence of this is that local government has an increasing responsibility to deal with the growing number of social, economic and environmental issues facing rural areas.

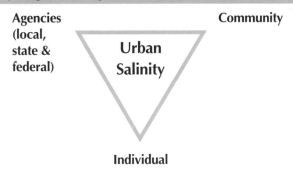

Figure 11.1 Urban salinity is the central problem addressed by the interaction of individual, the community and government agencies.

It is within this context that we document the way in which a shared community-government partnership has evolved in the rural city of Wagga Wagga in New South Wales. Wagga Wagga has high groundwater and associated salinity problems which have occurred as a result of a variety of land and water management changes during the last century (see below). We describe the maturing of the relationships between participants in the fight to rehabilitate areas of the city affected by salinity (Figure 11.1). From both an economic and social perspective the image of good clean country air and healthy lifestyles is discordant with those of a city which could in the future be labelled 'salt lake city'.

The salinity problem in the City of Wagga Wagga

The city's induced groundwater and salinity problem is known as dryland salinity because the problem has arisen in areas not irrigated for agriculture. Groundwater recharge in Wagga Wagga City is occurring through the following sources:

- over-watering on recreation areas, gardens and parks;

- over-watering by private individuals and groups of recreation areas and house gardens;

- leakage from the city's water supply;

- leakage from the city's sewerage system;

- drainage soak-aways; and

- rubble pits (porous drainage facilities constructed to remove surface water) (Bugden 1995).

The city's dryland salinity problem has features typical of irrigation induced salinity.

There is no fundamental difference in the geophysical processes involved in either dryland salinity or irrigation salinity; however measures to redress the problems differ. Typically, dryland salinity in an urban area is termed 'urban salinity' and we use the term to encompass the causal high watertable as well as the associated salinity. Salt (non-sodic) and a high watertable (waterlogging) result in different environmental impacts in urban areas (Eberbach 1998).

Salt (non-sodic) can cause the dissolving and recrystallisation of salts in mortar and brick through periodic wetting and drying, resulting in structural damage, and the corrosion of metal reinforcing in concrete. In addition, saline soils create water stress in plants, eventually leading to the death of some species, thus resulting in aesthetic problems in gardens.

High watertables (waterlogging) can cause damage to bricks and mortar from wetting and drying, resulting in bricks slumping over time. Waterlogging also leads to anaerobic conditions in soil, resulting in nutrient deficiencies and toxicities and the enhanced growth of soil pathogenic organisms which may promote disease in plant roots. The outcome is reduced aesthetic appeal in suburban gardens, though not as dramatic as that caused by salt.

The city of Wagga Wagga with a population of 56 000 people is estimated to have over 600 residential houses at risk of being affected by the rising saline groundwater. Because of the diverse nature of urban salinity and limited knowledge of its long-term effects, the extent of the economic impacts associated with the problem is difficult to establish. The costs of salinity can be separated into annual or recurring costs, and expenditure of public and private infrastructures that are mainly of a capital nature and not expected to recur regularly. Annual costs to 1995 were estimated to be over $400 000, while non-recurring capital costs were over $530 000 (Christiansen 1995).

The salinity problem in the Wagga Wagga City region is essentially a high watertable problem (Hamilton 1996). Hence, salinity control relies on two essential activities: reducing the amount of water getting into the watertable and removing water from the watertable. This is a technical analysis of 'the salinity problem' which has been identified by Sawtell and Bottomley (1989) in the Salinity Management Plan in the Shepparton Irrigation Region as leading to an over-emphasis on technical solutions, to the virtual exclusion of the social complexities of understanding and resolving salinity problems.

> 'While salinity is a high watertable problem, it is also a social problem with a 150 year history based on the economic, political, and social structures and values of European settlement in the region.' (Sawtell and Bottomley 1989, p. 3)

The urban salinity can be viewed as impacting at multiple levels: on government agencies, on the individual householder and on business.

Effects on local and State government infrastructure

In Wagga Wagga the effects of the high watertable and salinity are being felt on buildings, roads, sporting fields and golf courses, parks and gardens, water, gas and sewer supplies. The City Council continues to undertake repairs to underground infrastructure and road surfaces. Short (1997) estimates that roads are having to be reconstructed after 10 years rather than 20–30 years at costs up to $1 000 000 per kilometre; cast iron galvanised water supply pipes are corroding, sometimes only lasting 5 years instead of 50 years. Infrastructure repairs on a residential block basis are estimated to be approximately $300 000. To minimise the impact of urban salinity in the future the council is considering using higher cost materials during construction to better tolerate salinity and waterlogging, installing and repairing sub-surface drainage to alleviate the damage of high watertable, and other measures detailed later in the paper.

An Australian Bureau of Agricultural and Resource Economics (ABARE) survey of 177 councils in the Murray-Darling Basin in 1995 found 35% were spending money on repairs and maintenance believed to have been caused by salinity or rising watertables (Oliver et al. 1996). The estimated total repairs and maintenance over the 12-month accounting period was nearly $8.2 million.

Effects on households

Urban salinity is evident in the damage to brick foundations and bricks in the lower courses of houses, cracked and collapsed pavements, damage to driveways and flaking of internal walls. Approximately 100 houses require immediate work to repair damage, while 600 houses are at risk (Short 1997). Some residential homes installed electric pumps and tile drainage to reduce local groundwater problems at a cost of about $20 000; the practice is now illegal.

Other evidence of damage to residential buildings in Wagga Wagga as a result of the rising saline watertable includes:

- drops in domestic water supply pressure;
- soggy backyards where old rubble pits no longer function;
- corroded underground water, gas and sewer pipes;
- corroded metal fences;
- corroded outside metal items such as hot water systems and water fittings;
- some vegetation dying;
- good growth of salt-tolerant vegetation;
- damp course failing and dampness rising above the floor level;
- plaster falling off the walls of solid brick houses;
- external brickwork turning powdery; and
- breakdown of brick piers.

The social cost

The social cost of urban salinity is difficult to quantify, remains largely hidden and is likely to grow. Numerous families have sought loans (between $5000 and $20 000) in order to make repairs to their salt-damaged homes. The Landcare group has evidence that three banks will now no longer lend to those people seeking to buy a home in a known salt-affected area. Most people purchasing a home are unaware they may be purchasing a 'salt house' which has existing salt-induced damage or may have in the future.

While the problems associated with rising watertables are increasing, currently most of the salt-affected areas occur in the middle to cheaper housing areas. People in these areas are least able to take further loans to undertake salt repairs.

Conceptual problems

There are several features of urban salinity, relating to cause and effect, which impact on the degree to which the problem is perceived as serious by the community and political leaders. These features include:

- the time lag between action and impact;
- distance between the cause (recharged areas up slope) and the effect (discharge areas down slope);
- the idea that 'salinity is not a problem in my backyard';

- reliance on all levels of government to fix the problem; and

- lack of detailed biophysical information.

There has been a time lag between clearing of native vegetation, the construction of housing and other infrastructure and the impact of land-use change on the environment. The rise in watertables immediately following land-use changes has been quite small: in the vicinity of a metre or less. Because little is known about historical variation in the watertable in different parts of the city it has been difficult to establish the true rise in salinity over time. Thus members of the community advocating action on salinity issues are often unable to attract much support.

A second feature of the urban salinity problem, relating to cause and effect, is that a cause of the problem (the recharge area in Wagga) has usually been some distance from the effect of the problem (the discharge area). Impact sites are usually in the next suburb, down slope, a couple of kilometres away. Additionally, identification of recharge areas has been problematical, although work by the Wagga Wagga City Council and the NSW Department of Land and Water Conservation is currently in progress to map and document recharge, discharge and watertable levels throughout the city.

People may acknowledge that urban salinity is a general problem in the community. They often argue that it is not an issue that affects them, or which they contribute to. However an understanding of the salinity process (above) results in the recognition that salinity impacts on all of us; our rates and taxes are paying for maintenance of roads, and other infrastructure.

A fourth feature of the problem is an expectation that since the problem has such massive repercussions then government will step in and solve the problem at the end of the day. What is emerging however is that governments are continuing to cut back expenditure in the face of enormous pressures to meet basic human needs such as health and education; the concept prevalent is 'the user or abuser pays'. Compounding the reluctance to allocate taxpayer money is the knowledge that changing human practices at all levels — individual, community and government — is necessary to overcome the problem; no amount of money alone can solve the problem.

Political issues

Investigations by State and local government authorities and evidence collected by community groups indicate that the trend in rising watertables is continuing. In Wagga Wagga (Hamilton 1996) the community faces a problem that can at best only be partially dealt with by individuals, despite the availability of some feasible technical solutions. Thus political solutions are necessary. Our involvement in the urban salinity debate in Wagga Wagga caused us to reflect on the nature of political institutions. Some of the stumbling blocks to political solutions to issues such as urban salinity are generic; others are specific to smaller urban centres.

Generally, governments are under the influence of specific interest groups who vie for limited resources. Favouring of a few specific interest groups often occurs at the expense of projects that serve the general good. In contrast, common interest groups lobby for support of projects in the interest of the whole community. This has been well summarised by Etzioni (1993, p. 132):

> 'When Common Cause fights to reduce the power of private money in public life, it is not seeking payoffs for its members; it is advancing what its members and leaders believe is in the common interest. This sort of cause is quite discernible from say an oil company's argument that it should be given a tax deduction because oil drilling is in the national interest. The criterion to be applied is this: Who benefits? If it is

mainly the members of the group in question then it is a particular interest whatever the group's pronouncements dress it up to be. If the beneficiary is mainly the society at large then a common interest has been found.'

Advocates for action regarding urban salinity faced this general problem when seeking solutions from local government. Most such advocates have not come from the ranks of the more powerful business sector; rather their ranks include academics, a few farmers from near-city locations, students, retired persons and people with home duties. Their strategies to implement change until recently have not been political; rather their approach has been on-ground works such as tree plantings, speaking engagements and school education. As the group moves from naive conceptions of the political process to more sophisticated conceptions, so too do the strategies for change they employ.

A particular problem in smaller communities concerns the available human resources. Even at local government level the political process by its nature excludes large numbers of people, particularly those people who have large time commitments to work and family or carer duties. The process excludes those who have irregular working hours, such as shift workers. It excludes those who have already committed their available time to other community work. When surveyed many of the 15 councillors on the Wagga Wagga City Council allocated over 25 hours per week on council-related tasks. For several councillors this was in excess of 40 hours per week (Vidler 1997, personal communication).

Rural Australia is currently undergoing immense structural and social change and political processes are far from being egalitarian. Consequently many social, economic, cultural and political differences in the community do not input to local government policy. Many local governments have taken genuine measures to involve community representatives in committee forums. The *Local Government Act 1993* has also encouraged greater community participation.

Urban salinity, and the slowness of local government to move on this issue, is a particular case in point. If every person in Wagga Wagga City had been provided with evidence to show that one-quarter of the public and private buildings are in known salt-affected areas, and that salt is estimated to be costing the community at least $500 000 a year, then perhaps the public may have influenced council policy on urban salinity at an earlier date (see below).

Local responses to salinity

Awareness: starting to talk about it

Salinity became a tangible problem for residents of the area in the early 1990s. However there was a reluctance to acknowledge salinity as a problem for residential buildings. The issue was one that that you didn't talk about and if you did no one would believe you. Some people believed that their properties, if identified to be in salt-effected areas, would be virtually unsaleable or have values adversely effected. Others believed the city itself, and the publicity it would bring, would have an adverse effect on the prosperity of the region.

The experience parallels that in the farming district of Shepparton. The talking, however, continued as further evidence became inescapable. A survey of residents in a severely affected area in 1995 found only 41% of householders interviewed knew of a soil/water problem, while 50% said they did not have a problem and 9% stated they did not know if they had a problem in their area (Centre for Conservation Farming 1995). The business community were better informed with 67% stating they knew of a problem in their area.

A turning point for many residents came when the local agricultural show society lowered the inner ring of the showgrounds in 1989 by about 1 metre with the result that large salt scolds appeared on a very high profile site in Wagga Wagga. The catchment soil conservation officer identified the saline high watertable as the cause. Since then additional areas have been identified as salt-affected, including a once popular football oval which is now parched and unusable.

A decade later State government agencies, council and community groups are working on various groundwater control mechanisms, from technical solutions involving the installation of groundwater pumps to lower the watertable, to social solutions involving changed individual practices and community action such as tree planting, to the political, involving policy at local government level. But what caused this dramatic shift in attitude and commitment? This chapter argues that it was not the apparent evidence of saline watertables that caused the shift, but rather the broadening of the political power base. This occurred when the issue was adopted by the specific interest group whose members dominated the local council.

Responses of government

The State government and local council have acted singularly and together to identify and ameliorate urban salinity (see below). Reductions in available professional staff owing to government downsizing and the shift of funding to projects with community involvement has encouraged professionals to collaborate with the community. Additionally the community welcomes the technical advice, typically beyond its resources and capacities.

Government initiatives include:

* establishment of an Urban Salinity Working Party;
* development of a multi-agency land and water management plan;
* development of a groundwater 'map' by NSW Department of Land and Water Conservation as part of a land and water management plan for the city, identifying the extent of the problem;
* the public release of the groundwater 'map' with public consultation;
* installation of a pump out bore in the worst affected area to assess the effectiveness of lowering the watertable by this means;
* trialing a number of other water-recovery mechanisms;
* a trial demonstration site, replacing rubble pits with rear block stormwater drainage on six houses, to assess the cost of retrofitting stormwater drainage within the Wagga Wagga area;
* installation of computerised water management systems on the council's parks and gardens; and
* establishment of a Water Wise demonstration garden at the Botanic Gardens.

In June 1997, the Wagga Wagga City Council committed two and a quarter million dollars over 3 years in specified programs aimed at redressing urban salinity. These programs included on-ground works to lower the high watertables and the appointment of an urban salinity educator to promote awareness and encourage a change in water management practices.

The Wagga Wagga Urban Landcare Group

The group was formed in 1993 with a general aim of targeting environmental problems but with the primary aim of helping to redress the urban salinity problem through involving people with technical and research knowledge and the general community in action. That action involves physical on-ground works and community education. The emphasis is on working with people rather than on or for them. The philosophy of this Landcare group draws on many aspects of action research (eg Bawden and Packman 1991). This is because the rationale for the group's formation — redressing the salinity problem — was at the time, and still is, being investigated by biophysical scientists outside the social context of the urban community.

The problem itself is still being described. Identifying the problem involves local people, in particular members of the Urban Landcare Group, and researchers working together with the aim of learning from the experience and taking action to rectify the problem. In this context the researchers' knowledge about issues and problems is not assumed to be greater that that of the local people, and all participants feel they have some ownership of the information arising out of any collaborative efforts.

At the outset Landcare members battled a period of frustration. It was difficult to solicit more active members and there was a high workload on individuals to achieve on-ground works. There was also a failure to persuade political representatives of the seriousness of the problem and thus a failure to attract larger commitments from the city council, relative to the large allocations of council funds directed towards other projects. In addition, there was a rejection from those in the community who disagree with a public disclosure of the salinity issue, fearing the city will gain a reputation, deserved or otherwise, as 'salt lake city'. Finally, continuing technical debates about the precise causes of and remedies for the problem tended to give ammunition to the arguments of those who sought no action and no discussion on urban salinity.

The future

The Docent program — a joint community and agency effort

The Wagga Wagga City Council, the Department of Land and Water Conservation and the Wagga Wagga Urban Landcare Group are participating in an education program, the Docent program, funded under a Salt Action grant. ('Docent' is a term used to describe volunteers who are respected as leaders with proven abilities in various fields, and possessing highly developed community networks.) The primary aim of the Docent program is to inform the community on issues relating to urban salinity via the medium of nominated persons chosen for their relevant skills, knowledge and community contacts. Currently about one dozen people are acting as Docents. The Docents underwent a 2-month, 1-night-a-week training program provided by the agencies. A speaker's kit with video, literature and a set of photographic slides has been provided to participants who have been available for speaking engagements since February 1998.

The program was planned to coincide with the release of the salt-risk map which was expected to result in a high level of community anxiety, particularly from those householders designated to be in the current salt-risk areas. An increasing number of houses and commercial properties are expected to be threatened by salt as the watertable rises and moves progressively further up slope. Surprisingly the release of the salt-risk map did not result in any community outrage or indeed any perceived heightened anxiety as a result of the 'salt problem'.

The key aim of the Docent program is to significantly increase participation by the community in appropriate 'sound salt and water practices' as a result of the encouragement and education by the Docents. The target audiences for the Docents are the print and electronic media, community groups, the business and commercial sector, schools, government bodies and conservation and other groups already involved in urban salinity related activities.

Some of the sound salt and water practices the Docents are encouraging the community to adopt include to assist in the removal of all rubble pits; adopt sound water usage; ensure residential water mains are not leaking; and participate in public and private tree plantings. It is expected the some Docents will initiate study circles operating along lines similar to those of the Australian Association of Adult and Community Education (1996). The study circles do not follow a formal curriculum, but provide a forum for discussing social, environmental or political issues in an informal, friendly way, drawing on adult learning principles (Brookfield 1986), in order to increase knowledge and change attitudes relating to urban salinity. Members are encouraged to:

- listen to all views;
- debate the merits of those views;
- make choices about the subject; and
- taken action on the basis of knowledge.

The key messages that the community Docents and the Wagga Wagga Urban Landcare Group are seeking to convey are:

- We are all affected by the rising salt-laden watertables — our roads, underground and above-ground infrastructure and buildings are being damaged.
- We all have a stake in ensuring the viability of Wagga Wagga into the next century and beyond. There are things we can all do to help halt the rising watertables.

Making linkages

There appears to be a slowly growing appreciation that urban salinity has social, ethical, economic and ecological dimensions. Achieving a solution to sustainable management of salinity in Wagga Wagga will require a focus on all dimensions (Dorcey 1991).

The Wagga Wagga Urban Landcare Group and other groups have applied social and political levers at local government and State government levels, frequently filling the vacuum which resulted when government services and staff were withdrawn. In filling that vacuum the community has struck informal agreements with the remaining government personnel. Very often collaboration and joint participatory projects have evolved between the parties. The participatory approaches however do not always guarantee success or satisfaction, as inevitably each party functions with varying expectations and time frames.

Local government funding has served to cement the 'partnerships' and reinvigorate collaborative efforts. Successful participation has been contingent on participators and the community at large endorsing a philosophy of social co-operation. That philosophy has meant a shift from the assumption that the State will always provide towards an emphasis on joint voluntary and State action. Strategies being adopted in Wagga Wagga to redress urban salinity are being scrutinised by councils, other government agencies and Landcare groups around Australia as the problem of salinity is cemented in the Australian psyche as one of the country's great challenges.

References

Australian Association of Adult and Community Education Inc. (1996). *Guidelines for effective study circles.* (Australian Association of Adult and Community Education Inc.: Canberra.)

Bawden, R.J., and Packham, R.G. (1991). Improving agriculture through systemic action research. In *Dryland farming: a systems approach*. (Eds V. Squires and P.G. Tow.) pp. 261–270. (Sydney University Press: Sydney.)

Brookfield, S.D. (1986). *Understanding and facilitating adult learning: a comprehensive analysis of principles and effective practices.* (Jossey-Bass: San Francisco.)

Bugden, G. (1995). *Urban salinity: identification.* (Unpublished. Department of Land and Water Conservation.)

Centre for Conservation Farming (1995). *Urban salinity social survey: Wagga Wagga.* (Centre for Conservation Farming, Charles Sturt University: Wagga Wagga.)

Christiansen, G. (1995). *An economic report on the costs of urban salinity in the city of Wagga Wagga.* (Department of Land and Water Conservation: Wagga Wagga.)

Curtis, A.L. (1997). Landcare, stewardship and biodiversity conservation. In *Frontiers in ecology — building the links.* (Eds N. Klomp, and I. Lunt.) pp. 143–154. (Elsevier: Oxford.)

Dorcey, A. (1991). Conflict resolution in natural resources management: sustainable development and negotiation. In *Negotiating conflict resolution in Australian water management.* (Eds J.W. Handmer, A.H.J. Dorcey, and D.I. Smith.) pp. 92–122. (Centre for Resource and Environmental Studies, Australian National University: Canberra.)

Eberbach, P. L. (1998). Salt-affected soils: their causes, management and cost. In *Agriculture and the environmental imperative.* (Eds J.E. Pratley and A.I. Robertson.) pp. 70–97. (CSIRO Publishing, Melbourne.)

Etzioni, A. (1993). *Public policy in a new key.* (Transaction Publishers: New Brunswick, USA.)

Hamilton, S. (1996). *Urban salinity investigations in Wagga Wagga.* (NSW Department of Land and Water Conservation: Wagga Wagga.)

Martin, P., and Woodhill, J. (1995). Landcare in the balance: government roles and policy issues in sustaining rural environments. *Australian Journal of Environmental Management* 2, 173–183.

Oliver, M., Wilson, S., Gomboso, J., and Muller, T. (1996). *Costs of salinity to government agencies and public utilities in the Murray-Darling Basin. ABARE Research Report 96.2.* (Australian Bureau of Agricultural and Resource Economics: Canberra.)

Sawtell, J., and Bottomley, J. (1989). *The social impact of salinity in the Shire of Deakin and Waranga.* (Goulburn Regional Advisory Council: Goulburn.)

Short, B. (1997). *Is urban salinity a problem in Wagga?* (Wagga Wagga City Council: Wagga Wagga.)

Growing food and growing houses:
preserving agricultural land
on the fringes of cities

Ian W. Sinclair

Introduction

The preservation of Australia's agricultural land in the face of expanding towns and cities is becoming an issue that needs to be addressed. The issue is exemplified by land on the fringe of metropolitan Sydney. There are three facts that need to be addressed when dealing with this issue: the expansion of Sydney (or any town or city); maintaining agricultural production in the Sydney Basin; and dealing with rural land use conflict. All of these issues come under the heading of growth management and the ways in which we expand our cities. The issue is occurring to some degree throughout rural Australia; however it is more acute and apparent on the fringe of Sydney.

There is a need to provide a balance between the pressure of urban growth and the need to protect high quality agricultural land from further fragmentation and alienation. The first step is for agricultural land to be recognised as a legitimate land use planning constraint when considering growth management strategies for the expansion of Sydney and rural cities into the rural hinterland.

This chapter is based on experiences gained preparing planning policy for Wollondilly Shire which is located on the south-western fringe of Sydney. This region is experiencing a high rate of population growth which is related to urban growth pressures from Sydney. It also has a highly productive agricultural area. Wollondilly is also experiencing the relocation of agriculture from other urban growth areas of Sydney. In response to this, Wollondilly Shire has recognised the importance of this issue and has developed a land use zoning regime that places importance on the preservation of agricultural land. In order to provide for the protection of this land, specific land use planning policies have been adopted, including three new zones aimed at providing a balance between the need to protect agricultural land and the desire for rural living and limited urban expansion.

Growth management

Australia's agricultural land is a finite resource. Contrary to most beliefs, only 10% (70 million ha of 768 million ha) of Australia's land mass is arable land suitable for soil-based agriculture and livestock production; much of this is marginal with respect to water and nutrient regimes (Nix 1978). Most of this land is located on the coastal fringe of the continent. Australia is a heavily urbanised country. In 1991 82.5% of the population lived in urban centres of 10 000 or more (71.7% lived in urban cities larger than 80 000 people) (State of the Environment Advisory Council 1996, pp. 3–4). Most of these centres are located in the coastal or near coastal fringe of the continent between Brisbane and Adelaide.

As Australian towns and cities expand, they are expanding at the expense of the rural hinterland. Sydney is Australia's largest city with a population of 3.88 million people at 30 June 1996. This represented 62.5% of the total population of New South Wales. The population of Sydney increased by 60 000 (0.96%) between 1995 and 1996 (Australian Bureau of Statistics 1997, pp. 4–5). The fringe localities of the Sydney region are growing at a much higher rate. As Sydney's population grows there is pressure put on its rural hinterland to be converted from rural to urban use. This has been a common occurrence as once productive agricultural land is converted to residential land. This land use change is occurring at a relatively high rate on the urban fringe of Sydney. The pressure to develop land for urban development and rural residential development is growing as the Sydney region population continually increases. Topographically, Sydney is a basin. Urban expansion and the associated rural residential uses have reached the edge of the basin and are coming into conflict with agricultural and environmental issues on the fringe. Agriculture cannot move over the lip of the basin into the western areas because, as we saw in the recent drought, water availability cannot be guaranteed. Nor is there the workforce, and the cost of relocating the infrastructure is very high. The soils, generally, are not as fertile for the intensive nature of agricultural production west of the divide. If this urban expansion is to continue unchecked we will run out of productive, high quality agricultural land in the Sydney region.

Associated with the actual urban growth is a desire for residential living in a rural setting (known generally as rural residential development). This type of living opportunity, if not segregated, invariably creates conflict with the surrounding agricultural land uses which are being practised on the high quality agricultural land. This conflict arises when these rural residential lots are scattered throughout the countryside and next to productive farms. These incompatible land use patterns result in major conflict as is currently being seen with poultry operations and market gardens. This conflict, if left unchecked, will only intensify as agricultural production is forced to relocate from other parts of Sydney. The location of unplanned rural residential sites next to productive agricultural land is perhaps a more important issue for many parts of Australia beyond the metropolitan fringe.

Agriculture in the Sydney region

The protection of high quality agricultural land within the Sydney region is an important issue for the future planning of the region and its fresh food supply and is the main theme of this chapter. Sydney has a favourable climate with reasonable rainfall and a long growing season, as well as water available for irrigation, good alluvial soils and a readily useable workforce. This combination gives

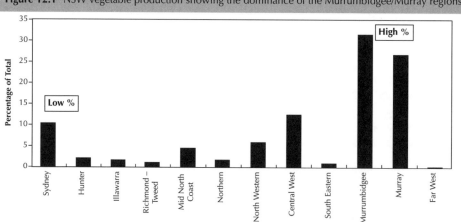

Figure 12.1 NSW vegetable production showing the dominance of the Murrumbidgee/Murray regions.

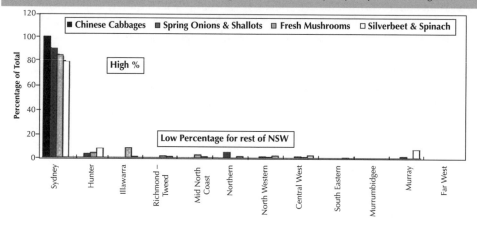

Figure 12.2 NSW vegetable production showing the dominance of Sydney for perishable vegetables.

it an advantage not found elsewhere in NSW. It is the shires located on the fringe of Sydney which produce a significant proportion of the fresh produce both consumed and produced in New South Wales (especially the perishable commodities). The main items grown in Sydney's urban fringe are perishable vegetables, poultry, nursery products, flowers and cultivated turf.

Australian Bureau of Statistics data for the 1993–94 Agricultural Census has been analysed to illustrate this point (Sinclair 1997). Vegetable production occurs in all regions of NSW. The Murray and Murrumbidgee regions produce the highest percentages of total vegetable production (27% and 32% respectively). The Sydney region contributes 10% of the total vegetable tonnage produced in NSW (see Figure 12.1). However, when one breaks down this vegetable production into perishable commodities (those commodities that perish quickly when harvested and therefore need to be located close to the market they serve), a different picture arises. It can be seen from Figure 12.2 that the Sydney region produces 52% of the State's lettuce production 84% of fresh mushrooms, 78% of spinach, 90% of spring onions and shallots, 98% of parsley and 100% of the total tonnage of Chinese cabbages. The Sydney region also accounts for 40% of the State's total area devoted to nurseries, 53%

Figure 12.3 NSW nurseries, flowers and cultivated turf showing the dominance of the Sydney region.

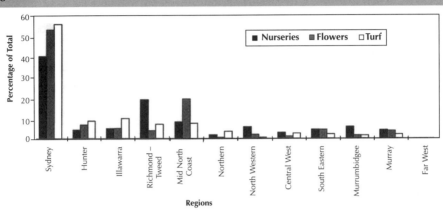

Figure 12.4 NSW poultry production showing the dominance of the Sydney region.

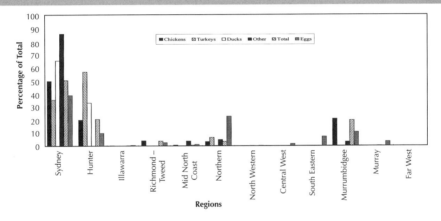

of flower production and 56% of the total area under cultivated turf (see Figure 12.3). Chicken production in the Sydney region accounts for 51% of the State total. Figure 12.4 illustrates this graphically; it also shows that the Sydney region is where the various types of poultry products (chickens, ducks and turkeys for meat as well as eggs) originate. NSW Agriculture estimates the value of agriculture in the Sydney region to be $1 billion (NSW Agriculture 1998, p. 14).

There is a high proportion of high quality agricultural land in the Sydney region. In fact, approximately 20% of the land area left is classed as prime agricultural land (Classes 1–3).

The retention of the broad range of agricultural production in the Sydney region generates a wide range of benefits, the main ones being its economic contribution, its market proximity and its ability to provide a land use that is compatible with catchment management for the Nepean Hawkesbury River.

In addition to the real dollar value of agriculture mentioned above, it also has economic benefits for the Sydney region. The agricultural industry can generate substantial economic linkages and multiplier effects to other industries such as retailing and other commercial uses (banking, solicitors, accountants, etc.). The multipliers for agriculture vary significantly but are generally recognised to be in the order of two to three. That is, every dollar of agricultural output is worth $2 or $3 to the economy in total as a result of flow-on affects. However, this multiplier may not include the subsequent processing of agricultural products. When such value added processes are included the total value can be two to three times the initial multiplier. In the case of the poultry industry the economic multiplier effects have been identified as being 6.77 for employment, 3.25 for output and 0.58 for household income (Larkin and Associates 1991).

Employment in the agricultural sector can be broken into two components: full-time and part-time. This is a result of the generally intensive nature of agriculture practised in the Sydney region. There is a large part-time seasonal workforce which can be supplied from the large adjoining population. This workforce is employed in the picking of fruit and the harvesting of vegetables and other produce — all very seasonal activities requiring a pool of flexible part-time workers.

Sydney's large population means that it is a large market for fresh produce. The fragility and perishability of much of the fresh produce make it essential to have quick access to the markets so that it reaches the consumers fresh. This market proximity means that farmers can take their own products to the markets and thus do not need to employ agents or carriers for market delivery. This allows for economies of scale. Market proximity for raw materials is also important for such uses as poultry where feed is produced in the Sydney region, as well as for the sale of the birds. It should be noted that most of the poultry processing plants are located in the Sydney region.

The continuing existence of agricultural development on the fringe of the Sydney region will maintain the rural character of this area. This has the benefit of providing an area which is close to metropolitan Sydney and allows residents to enjoy the rural landscape character. Such an experience is vital for the tourism potential as well as contributing to the quality of life of the residents of urban Sydney.

Another feature of regional agriculture is its highly interdependent nature. The poultry industry and market gardens are good examples of this. The market gardens utilise the poultry manure as fertiliser. Poultry also utilises wood shavings for the floors of the sheds. Thus there are benefits for the integration of different agricultural practices. The wastes from one practice become the raw materials for another. The mushroom industry is another good example. Mushrooms are grown in compost that is made from a mixture of straw (from wheat stubble and stable beds sourced from the racing stables in Sydney), cotton husks and poultry manure. Once the mushrooms have been grown the compost is sold as garden compost. This linkage is broken when one industry is forced to relocate to another geographic area.

The disposal of urban wastes, especially sewage, has major implications for the continuation of agriculture. The acceptable disposal of sewage is one of the major environmental issues facing the community at this time. Investigations are being carried out into the possibility of land disposal using agriculture and/or agroforestry practices. A sewerage treatment plant using this system is to be constructed in Picton serving the towns of Picton, Tahmoor and Thirlmere. Similar schemes are being proposed by the NSW Government in other parts of the metropolitan fringe and thus significant areas of agricultural land should be preserved for this purpose.

The Sydney region has soil characteristics, a growing season and rainfall patterns which are all suitable for the growing of most temperate crops such as deciduous fruits (citrus and pome fruits), fresh vegetables, cut flowers and nursery stock, as well as cultivated turf.

The agricultural land uses also have a considerable medium-term capital investment associated with them. This includes buildings and structures such as poultry sheds, packing sheds, machinery sheds and hot houses ('igloos') as well as hail and bird netting. Plant and machinery used on the properties are also included here. An example of the magnitude of this investment can be found in the case of poultry sheds in Wollondilly. The cost of building a new poultry shed today is in the order of $150 000 per shed. The Wollondilly Agricultural Lands Study found that there are a total of 277 poultry sheds in the shire. This equates to an investment of $41.5 million in today's dollars. The existing investment in agricultural infrastructure, therefore, is a reason for the continuation of agricultural production given the high cost of replacement. It stands to reason to use what is there at present than to go through the costly procedure of relocating agriculture.

Rural land use conflict

The presence of agriculture and non-rural land use in the one location can often generate conflict due to their potential incompatibility. Agriculture can affect adjoining small rural lots which are used essentially for residential purposes. Similarly, the presence of small rural lots creates an adverse influence on the continued operation of agriculture. The issue of rural–urban conflict arises when there is no separation between incompatible uses, let alone the misunderstanding which may exist about the purpose and character of a district. Land use conflicts may arise in such situations through noise, odour, farm chemicals, light, visual amenity, dogs, stock damage and weed infestation, to name just a few.

When considering these conflict issues it is important to remember that agriculture is a dynamic activity utilising a range of practices and equipment commonly unfamiliar to urban-based people. The notion of a rural lifestyle is engendered by an association with the pleasant character of the landscape rather than the potentially offensive noises, odours and operations which are the reality in the agricultural areas of a rural shire. Increasing competition for the available land tends to intensify the agricultural practices at a particular site, thereby increasing the potential for conflict with non-rural residents.

It is such a picture which characterises agriculture on the fringe of the Sydney region. Much of it is intensive by nature, given the typically small property size in preferred agricultural areas. Vegetable growing, turf farming, flower growing, nurseries and poultry production are important, along with horses, cattle grazing, dairying and orcharding. All of these agricultural enterprises are able to benefit from the market advantage gained by being so close to Sydney.

There are a large number of rural small holdings offering rural residential living or hobby farming scattered throughout the rural areas of fringe shires and other parts of rural Australia. Many are generally 2 hectare allotments which have been excised from the larger adjacent holding with no thought of the implications of this or the future conflict that will occur. There is also a significant proportion of lot sizes up to 10 ha that are used predominantly for rural residential use (Wollondilly Shire Council 1993).

The resolution of this conflict in the use, attitude and perception of the rural zone is difficult to reach because of its complexity. It will not be easy especially since each 'player' (farmer and rural-resident) has different ideas about the use of their land. Neither lives in a vacuum; they live within a community and they have the opportunity to exercise their rights and responsibilities, as well as the ability to influence future decisions about their local area. It must be remembered that people need to eat. The resources to provide this food and fibre are not unlimited and the longevity of the resources depends upon the sustainable use. Similarly, people have a right to live, but in a manner which does not compromise the existing, and possibly necessary, use of the land, while remembering that they have certain obligations as responsible community members.

Certainly education about the issues of agriculture and conflict with adjoining properties is fundamental to the resolution of the conflict over land use. How many residents, whether rural or urban, know that the Sydney region produces a high percentage of NSW's perishable vegetables, poultry, nurseries, cultivated turf and cut flowers? Do they wish to compromise this essential component of their local economy through unfounded attitudes on the future use of the rural land in their council area? The dissemination of information about agriculture, combined with growing awareness of the scarcity of sustainable land resources, will enable all residents to make more informed decisions about the desirability of one land use compared to another in a particular location.

There is a need to separate incompatible land uses while recognising the efficiencies which can be achieved through the integration of many of these land uses. This may be achieved, for example, through physical separation or a simple vegetative buffer designed to screen one land use from another. Land use zoning can also be used and this will be discussed in detail later. Such practical strategies require those with potentially conflicting land uses to acknowledge their impact and then design their operations to account for this impact. A community approach utilising physical solutions, planning strategies and a long-term vision for the land use of the shire will enable sustainable coexistence of agriculture and residential land use.

Local government can play a further role in the overall education process. Many rural Councils for example, place a special notation on all Section 149 certificates (these are required to be part of a contract to sell land and stipulate the zoning and other planning issues associated with a parcel of land) so that all future purchases of rural land, especially rural residential purchasers, are advised of the surrounding agricultural uses and thereby the potential for conflict. This can be combined with media releases and other methods of disseminating the agricultural message to inform this diverse audience, such as articles in council newsletters and the Annual Report. Other government departments can also assist. NSW Agriculture, for example, produces farmer publications (Agfacts, producer newsletters) that can aid in educating the public.

The case of Wollondilly

Wollondilly Shire is located on the south-western outskirts of the Sydney region. It has a population at 30 December 1997 of 34 600 people. Wollondilly has experienced an annual average growth rate between 1986 and 1991 of 4.3% per annum. This was the second highest growth rate in the Sydney region. The current annual growth rate is 2.2%.

Wollondilly is characterised by scattered settlements with 16 towns and villages ranging in size from Nattai with 75 people to Tahmoor with 4000. Wollondilly has a demographic make up which is characterised by a high proportion of the population under 19 years (36.6%) and a high proportion of the population in the 30–49 year age category (32%). Only 13.4% of the population is aged 20–29 and 18% of the population is over 50 years. It can be seen therefore that Wollondilly is an area which has a high proportion of young people. It is also expected to take a proportion of Sydney's growth over the next 20 years.

The major issue for the future development of Wollondilly is the pressure for urban growth of the Sydney region. Associated with this is the role of agriculture as well as environmental problems with water quality in the Nepean Hawkesbury River and air quality of the Sydney region. Wollondilly is in the headwaters of the Nepean river and people involved in any development will have to be mindful of the potential for an adverse impact on the Nepean Hawkesbury River downstream of the shire. Wollondilly is a high producer of fresh vegetables, nurseries, flowers and cultivated turf as well as poultry. The value of agriculture is estimated at $360 million (Kelleher *et al.* 1997) Wollondilly also has 54% of its privately owned land classified as high class agricultural land.

Analysis of the Australian Bureau of Statistics Agriculture Census for 1993–94 shows that Wollondilly is a particularly high producer for both the Sydney region and NSW for certain vegetables and poultry. It is one of the highest poultry producing shires in the State and is the fourth highest producer of vegetables in Sydney. Wollondilly is obviously an important contributor to agricultural production in the Sydney region. Its location and large size mean that it will continue to play a role in agricultural production in the future as long as the value of agricultural land use is reflected in strategic planning decisions for the fringe of Sydney.

Planning for balanced urban and rural development is currently being addressed by Wollondilly Shire. Council is currently undertaking planning studies to address the issues of urban growth, environmental impact and the role of agriculture in the shire and the Sydney region. It is felt that planning is the only way that all of these competing interests can be accommodated to ensure the future development of the Sydney region. We are currently at a watershed in the planning for the future of Sydney and if some drastic planning measures are not taken soon the results could be disastrous for the whole Sydney region.

Planning for the preservation of agricultural land

Planning for the preservation of agricultural land is an important issue that has not been adequately addressed by governments at all levels. Some councils have addressed the issue in detail. It is proposed to discuss this issue by using the Wollondilly Shire as a case study. Wollondilly Council has been developing a detailed land use policy to protect its agricultural land for the past 6 years. It started with the Wollondilly Agricultural Lands Study, published in 1993 and was followed by the Review of Rural Lands Local Environmental Plan (LEP) and Development Control Plans (DCPs).

Wollondilly Agricultural Lands Study

The first stage in the preparation of any planning exercise for rural land is to have a detailed understanding of the existing situation. The *Wollondilly Agricultural Lands Study* was prepared by the council to provide that detail. The study was published by Council in April 1993. The catalyst for its preparation was a concern about the loss of agricultural land and the issues of rural land use conflict.

One of the most important components of the study was the advisory group that was set up to help to prepare it. This group had a wide ranging representation. All of the main categories of agriculture in the shire were represented (extensive and intensive livestock and horticulture) as well as government departments and councillors. This group was used as a sounding board and also to check the accuracy of the data presented. The existence of this group was instrumental in the acceptance of the study and its recommendations.

The study was a detailed analysis of agriculture in Wollondilly as well as the Sydney region. It provided an analysis of agricultural production in the Sydney region as well as in NSW. Rural land use conflict was identified as an issue and its causes were discussed. A detailed analysis of rural land use was carried out and the results of this were discussed, along with the implications of the change in agricultural land use. (This included the benefits and threats to agriculture in Wollondilly as well as in Sydney.)

The study was based on a detailed land use survey. This survey has been separately published and identified all land uses within the rural area of Wollondilly. The survey categorised agriculture into four broad categories: intensive livestock and horticulture and extensive livestock and horticulture. Rural residential land uses were identified, as were natural vegetation and other uses. In tandem with the land use survey, a detailed lot size analysis was carried out. The land use for each parcel was entered into Council's computer-based property system which allowed for cross-referencing of land use and lot size, useful for analysis purposes. The results of this are shown graphically in Figures 12.5 and 12.6.

The study highlighted the need to plan for the retention of land for agricultural production. The issues to be considered for the retention of agricultural land were highlighted and a methodology was developed for the assessment of land to be zoned for agricultural production. A discussion on the benefits of zoning for agricultural production was undertaken.

The study concluded that agriculture is an important land use in the Sydney region and as such should be recognised as a legitimate land use planning constraint when preparing strategies for the future. If agriculture is not recognised as a constraint to further urban development there will be major implications for the availability of fresh food and produce in Sydney. As a result of rural land use conflict, there is a need to identify specific zones for agricultural production and rural living areas (rural residential). The cost of not protecting agriculture for the future will not only be the loss of fresh food production in close proximity to Sydney, but also the loss of the rural hinterland and the benefits of this for recreation, tourism and water quality.

The recommendations of the study were to request the Department of Planning to prepare a Regional Environmental Plan for agriculture in the Sydney region and to recognise it as a legitimate land use planning constraint. Recommendations were also made for a review of Council's rural land to provide for agricultural production and rural living zones as well as requiring development consent for intensive agriculture and the preparation of a Development Control Plan to provide guidelines. The study also recommended that an education campaign be undertaken to highlight the benefits of agriculture and the issues concerning rural land use conflict. Finally, the study recommended that the council review its farmland rating policy. The implementation of some of these recommendations is discussed later.

Issues to consider

Planning for the preservation of agricultural land is a complex issue. One cannot consider any of the issues in isolation— they need to be considered together. There is a need to consider the range of lot sizes, land use, land suitability, locational factors, the critical mass of agricultural land and other issues related to agriculture. All of these must be considered when developing planning policies for the retention of agricultural land. Most importantly, there must be a detailed land use survey and lot size analysis so that a clear understanding is gained of the location, holding size and spatial distribution of the rural land uses.

Methodology for protecting high quality agricultural land

The methodology utilised for assessing land for agricultural production was developed by the Wollondilly Agricultural Lands Study. It takes into consideration issues such as lot size, land use, physical constraints, locational factors and rural land use conflict. The rationale behind the methodology is to reduce the incidence of rural land use conflict, thereby protecting the sustainability of agricultural development. It will also provide a balance between the use of the land for agriculture and the desire for rural living. It is shown graphically in Figure 12.7. It has since been used by other councils on the fringe of Sydney for their planning policies.

Planning for agriculture is an integrated process — all aspects are linked and one cannot be considered without looking at the impacts on all other aspects. Therefore, when applying the methodology it is necessary to consider all of the steps. A spatial integration of agricultural land use must also be considered. Planning for one farm cannot be considered in isolation from other land in the locality and other localities in the shire. Thus, it is important to take a holistic approach when planning for agricultural land use.

A methodology such as this is evolutionary because of the dynamic nature of land use planning. It has been developed for use in Wollondilly and has been applied in the review of rural lands recently undertaken. It is expected that it will be subject to scrutiny by planners and other persons and adapted to other areas. However, it has been developed by Council taking into consideration all aspects of land use planning theory and practice.

The methodology basically eliminates land which is not suitable for agricultural production. It is reliant upon a detailed analysis of the characteristics of agriculture within the locality such as that provided in the Wollondilly Agricultural Lands Study. The details are placed on a map to allow for ease of identification of all issues.

This methodology will lead to the identification of land which is suitable for and which should be retained for agriculture. This land falls into two categories: firstly, land that is currently used for and has good soil for intensive forms of agriculture such as market gardening, orcharding, grape growing, turf framing or poultry; and secondly, land that is not of as high a quality or productive yield and is generally used for extensive forms of agriculture. The former land should be zoned as agriculture and the latter as a mixed-use agriculture or agricultural landscape. It must be recognised that agriculture is the use and rural is the character. The methodology will also identify land that is more suited to rural residential use. It is important to base the zone boundaries on a catchment basis and use physical boundaries for the zones. The assessment should also be undertaken in consultation with NSW Agriculture and other government departments and organisations which have an interest in agriculture. Land which falls outside this classification can be considered for other uses, such as rural living, future urban or rural urban fringe development.

Review of rural lands — the Wollondilly solution

Wollondilly Shire has reviewed its rural land with the view to protecting productive agricultural land by providing for a balance between agriculture and the desire for rural living opportunities, as well as reducing land use conflict. The review has culminated in the preparation of a specific local environmental plan (LEP)and three supporting development control plans — one for the agriculture zone, one for the agricultural landscape zone and the other for the rural living zone. The development control plans are objective-based documents which will allow for the maximum flexibility in implementation. They also have a special section which explains the issues and reasons for implementing the specific controls. The LEP was gazetted on 20 September 1996.

Three new zones have been developed as follows:

- agriculture;

- agricultural landscape;

- environmental protection — rural living.

An important factor in delineating the zone boundaries has been the use of physical boundaries rather than a road or cadastral (lot) boundary as occurs in many urban zones. This is especially so for the rural living and agriculture zones. This has been done to reduce the incidence of rural land use conflict mentioned earlier. In fact, there are rural residential uses in the agriculture zone; however, the philosophy has been taken that it is not desirable to allow further subdivision on the edge of the production areas as this will lead to conflict. Also, intensive horticulture has been defined and requires consent in all agricultural zones and is prohibited in the rural living and other rural residential and urban zones.

It is felt that, by providing for agriculture to be located on suitable land, the future development scenario for Wollondilly can be achieved along the lines espoused by the council in its growth management planning, lines which do not encourage the 'aesthetic vandalism' of urban sprawl and instead promote planned sustainable communities.

Agriculture zone

There are five relatively small parts of the shire included in this zone. The primary objective of the agriculture zone is, as its title suggests, to preserve existing agricultural production and to allow for new agricultural production in appropriate locations by recognising the importance of high quality agricultural land.

The methodology discussed above has been applied to outline the parameters to be considered and the constraints which need to be recognised when assessing and identifying land suitable for an agricultural production zone. This land is generally high class agricultural land, known as prime crop and pastoral land. This includes classes 1–3 of the agricultural land classification technique used by NSW Agriculture as described in the *Rural land evaluation manual* (NSW Department of Planning 1988). However, some of the production areas contain Class 4 land which may be needed in a particular agricultural system.

A secondary objective of the agriculture zone is to reduce the incidence of rural land use conflict. This is done by ensuring that any new dwelling houses within the zone are permitted only in conjunction with a legitimate and sustainable agricultural enterprise. The impact of the dwelling on adjoining agricultural land is of prime consideration and the dwelling may have to be located away from the boundary or screening and mounding put up to reduce any potential conflict. Likewise, any agricultural enterprise proposed for land adjacent to an existing dwelling house which is used for residential purposes will have to take into consideration the use of the land for residential purposes and provide steps to reduce potential conflict.

There is a minimum subdivision size of 20 hectares within the agriculture zone. The proponent must satisfy the council that the subdivision is required for a sustainable agricultural purpose and as such a total farm management assessment is required. This includes a property plan, an agricultural sustainabiliity assessment and a farm management plan. Sustainable agriculture has been defined in the LEP as having three components: environmental (it minimises environmental pollution and or land degradation), social (it minimises land use conflict and loss of amenity for the surrounding area) and economic (it is capable of making a net farm profit).

Agricultural landscape zone

This zone covers most of the shire. The primary objective of this zone is to preserve the agricultural landscape of Wollondilly while also providing for agricultural production. It is this landscape character that gives the shire its distinct character and attractiveness for rural living opportunities.

Existing productive agricultural enterprises will be encouraged to continue within this zone; however, it is not anticipated that a great deal of intensive agriculture will be carried out. The areas set aside for this zone are generally areas which have a fair amount of extensive agriculture practised in the form of grazing for beef cattle and dairying as well as a predominance of deer farming. The minimum subdivision size for this zone is 40 hectares. This minimum is based on the historical minimum as well as on an assessment of the existing fragmentation within Wollondilly and the desire not to have any further fragmentation of agricultural holdings.

The issue of providing for small scale subdivision in conjunction with agricultural development has been addressed in this zone. The LEP makes provision for this to occur only under a community titles subdivision. This community titles subdivision must first of all be based on a total farm management assessment, mentioned above. A detailed landscape assessment must be provided and site constraint plan prepared; these will identify those lands which are suitable for dwelling houses. It is only then that the subdivision pattern may be provided. The minimum area for subdivision under the community titles is between 1 and 2 hectares; however, it must be pointed out that,

Figure 12.5 Wollondilly lot size analysis showing the majority of lots less than 18 hectares.

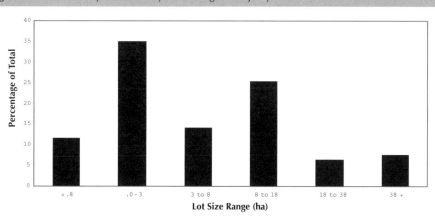

taking into consideration the constraints and specifics of the land, this may not be an achievable minimum in all cases. A maximum of 10 lots is permitted on any one holding.

Environmental protection rural living zone

Before describing the zone, it is first necessary to define the two different types of rural residential development within Wollondilly. The definition of each style of development areas follows:

- **Rural urban fringe** development is that style of development which is within the servicing catchments and in close proximity to an urban centre. It may have reticulated water and in fact may have reticulated sewerage although most effluent disposal will be on site. It will also have a garbage service. The lot size is generally in the range of 4000 square metres to 1 hectare and it is in an 'estate' style of development.

- **Rural living** development is a residential use of the land within a rural environment. It is not necessarily near a existing urban centre and does not have reticulated water or any other form of service which would generally be provided in an rural residential zone or urban centre. The density in Wollondilly is one dwelling per 4 hectares with a minimum of 2 hectares and any subdivision must be carried out having due regard to the constraints of the land.

There are five segments of the shire that are covered by this proposed zone and they predominantly have a residential use and varying lot size up to 10 hectares and above. As the title suggests, the purpose of this zone is to provide rural living opportunities within an environment which is sensitive. In fact, the whole of Wollondilly can be regarded as sensitive, given the issues of land degradation and water quality which must be considered, especially when subdivision to a smaller minimum is permitted.

The primary objective of the rural living zone is to provide for rural living opportunities having due regard to the preservation of the landscape character as well as the constraints of the land. Within this zone the methodology for carrying out a subdivision is first of all to identify the constraints of the land and then identify dwelling house sites. The actual subdivision layout is the

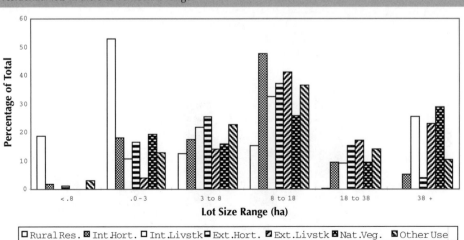

Figure 12.6 Wollondilly lot size cross referenced to land use. Note the high percentage of rural residential lots in the 8 to 18 hectare range.

last thing to be considered and in fact, having regard to the constraints of slope, soil quality and runoff potential as well as effluent disposal areas, the density of one dwelling per 4 hectares may not be achieved in many cases.

Conclusion

The preservation of agricultural land is an important issue facing Australia today. The statistics mentioned about the arable agricultural land and the population density present a dilemma that needs to be addressed when planning for the growth management of the cities and towns. This is especially critical for the Sydney region. There is a need for the issue to be addressed by both the State and local governments in concert with each other.

Figure 12.7 Methodology for assessing land for agricultural production.

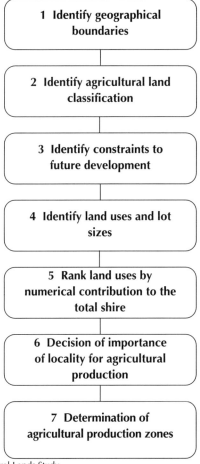

1 Identify geographical boundaries

2 Identify agricultural land classification

3 Identify constraints to future development

4 Identify land uses and lot sizes

5 Rank land uses by numerical contribution to the total shire

6 Decision of importance of locality for agricultural production

7 Determination of agricultural production zones

Source: Wollondilly Agricultural Lands Study

There is a need to provide a balance between the desires of rural residential living opportunities and maintaining long-term agricultural production. It requires knowledge of the resource base, limitations on its use, effective recognition of its environmental planning legitimacy and a creative approach to accommodation of the various land use requirements as proposed by Wollondilly Shire Council. Above all there must be a commitment to the process along with a clearly anticipated strategic planning vision for each fringe local government area, the region, and the State of New South Wales.

The provision of a balanced approach to rural land will allow for planned development which is economically viable, environmentally responsible, aesthetically pleasing and socially acceptable. It will allow for the growing of food as well as the growing of houses.

If agricultural land is not considered as a land use planning constraint we could run the risk of not having a rural Australia but a rural residential Australia after 2000.

References

Australian Bureau of Statistics. *Agricultural census — Season 1993–94, NSW*. (Unpublished data.)

Australian Bureau of Statistics (1997). *NSW regional statistics*. (Australian Bureau of Statistics: Canberra.)

Larkin, J.T. and Associates (1991). *The Australian poultry industry*. (NSW Chicken Growers Association: Sydney.)

Kelleher, F.M., Chant, J.J., and Johnston, N.L. (1997). *Rural subdivision — impact on agricultural industries, the natural resource base and socio-economic development of peri-urban environments*. (Rural Industries Research and Development Corporation: Canberra.)

Malcolm, B., Sale, P., and Egan, A. (1996). *Agriculture in Australia — an introduction*. (Oxford University Press: Melbourne.)

NSW Agriculture (1995). *Sustainable agriculture in the Sydney basin. Issues Paper*. (NSW Agriculture: Orange.)

NSW Agriculture (1998). Strategic Plan for Sustainable Agriculture — Sydney Region (NSW Agriculture: Orange.)

NSW Department of Planning (1988). *Rural land evaluation manual — a manual for conducting a rural land evaluation exercise at the local planning level, revised edition*. (Department of Planning, Sydney.)

Nix, N. (1978). Australia's natural resources. In *How many more Australians?* (Eds L.H. Day, and D.T. Rowland.) (Longman: Cheshire.)

State of the Environment Advisory Council (1996). *Australia: State of the Environment*. (CSIRO Publishing: Melbourne.)

Sinclair, I.W. (1997). Agricultural land is a constraint to growth management. In *Conference Proceedings — The changing agenda for planning*. (Ed. A.W. Witherby.) (Royal Australian Planning Institute.)

Wollondilly Shire Council (1993). *Wollondilly Agricultural Lands Study*. (Wollondilly Shire Council: Picton.)

Wollondilly Shire Council (1995). *Development control plan — rural living*. (Wollondilly Shire Council: Picton.)

Wollondilly Shire Council (1996). *Review of rural lands report*. (Wollondilly Shire Council: Picton.)

Wollondilly Shire Council (1998a). Agriculture 1(a) Zone Development Control Plan (Wollondilly Shire Council: Picton.)

Wollondilly Shire council (1998b). Agricultural Landscape 1(b) Zone evelopment Control Plan (Wollondilly Shire Council: Picton.)

Wollondilly Shire Council (1998c). Wollondilly Growth Management Strategy — A Growth Management Concept (Wollondilly Shire Council: Picton.)

Social and economic costs and benefits of taking water from our rivers: the Macquarie Marshes as a test case

Richard T. Kingsford

Introduction

The term 'wetlands' describes a range of freshwater ecosystems. Lakes, swamps, billabongs, floodplains and sometimes even rivers and creeks can be wetlands. Long regarded as sites of mosquito infestation and useless land, many wetlands were destroyed as Australia was colonised, and agriculture drained wetland areas and changed the flows of rivers for irrigation and hydroelectricity (Goodrick 1970; Norman and Corrick 1988; Halse 1989; Blackman *et al.* 1996).

Wetlands are now recognised for their biological importance around the world (Finlayson and Moser 1991), as well as in Australia (Australian Nature Conservation Agency 1996). Most people in Australia know of the wetlands of Kakadu even if they haven't seen them. Their importance to economies is becoming increasingly well known (Morrison and Kingsford 1997). Despite a growing recognition that wetlands are important ecosystems, there is considerable evidence around the world that they are under threat (Allan and Flecker 1993). They are disappearing at a significant rate. Many countries have lost large proportions of their wetlands (Micklin 1988; Hollis 1992; Weins *et al.* 1993; Jones *et al.* 1995; Anderson 1996). Australia may be no exception. Despite our small population, we have had a significant impact on the wetlands of our continent, particularly where agriculture and cities have become established (see McComb and Lake 1988). One of the key natural resources for agriculture is water and it is arguably the one in shortest supply for our dry continent.

Between 1994 and 1996, New South Wales went through the difficult process of revising a water management plan for the Macquarie Marshes where central issues of water management were debated at some length within the community. Access to water resources became a fierce public debate. The process revealed some important aspects of river management and some important principles for determining priorities for wetland conservation and protection of riverine ecosystems. The ecological impacts resulting from water diversion resulted in a change of water management which was not accepted by all in the community. Nature conservation might be a worthwhile concept except when it affects your business.

In this chapter, I review the main impacts that taking water from our rivers has had on the sustainability of wetlands. This has had not only significant ecological impacts but agricultural and social impacts on parts of rural Australia. There have also been benefits in terms of the establishment of rural communities. I focus on a particular wetland, the Macquarie Marshes, which has been the subject of considerable public debate. The aim is not to detail the level of degradation in the Macquarie Marshes — that has been done elsewhere (see Kingsford and Thomas 1995) — but to document community reaction to the policies which dealt with the wetland loss. Understanding different viewpoints is the first step towards reconciling these difficult resource management issues. There is no published compilation or analysis of the community's reaction to the debate on the Macquarie Marshes, but the contentious nature of the issue ensured that there was considerable public comment in the media (print, radio and television) which provides the material for such analysis.

Major causes of wetland loss or degradation in Australia

The functions of wetlands cannot be maintained if natural flow patterns are affected. As Australia was colonised, swamps were drained to increase agricultural land and reduce its flooding or to make way for urban centres (Goodrick 1970; McComb and Lake 1988). This was particularly true along the coastal fringe (Goodrick 1970; Pressey 1989; Blackman et al. 1996). On the lower floodplain of the Macleay River, 96% is affected by drainage (Pressey 1989). More generally, flood mitigation had reduced or destroyed about 60% of good waterfowl habitat on the north coast of New South Wales by 1969 (Goodrick 1970).

The next significant impact on wetlands was the regulation of our rivers with dams and weirs which control the flow of rivers to provide water primarily for human use. Dams are built to provide hydro-electricity (Snowy Mountains Scheme), to cool coal-fired power stations, to provide a water supply for drinking (eg Warragamba Dam near Sydney) and for industry, and to supply water for irrigation (Kingsford 1995a). Irrigation in Australia uses 70% of all developed water (Wasson et al. 1996) and over 95% of water in the Murray-Darling Basin (Murray-Darling Basin Ministerial Council 1995).

Between the 1950s and 1970s, there was a significant dam building phase as governments moved into 'conserving' water which was generally regarded as going to waste in wetlands or out to sea. Australia's storage capacity in major reservoirs totals 81 000 gigalitres (Wasson et al. 1996). In New South Wales, about 144 large (storage capacity greater than 1000 ML) dams were constructed, mostly in the headwaters of the Murray-Darling Basin and along the coast (Kingsford 1995a). There are more than 3500 licensed weirs on the rivers and creeks of New South Wales (Environment Protection Authority 1997). Weirs ultimately serve the same purpose as dams. They regulate flow for later use.

In principle, building dams and weirs should have little long-term effect on a river. Certainly as the dam fills, it deprives the river of water. All dams fill and then spill and in theory the river could behave as it did before the dam was built. But dams are built so that we can divert water for irrigation and urban water supply. Less water in a river means less flooding. One specific purpose of dams is to stop flooding. For example, a flood mitigation zone or airspace was operated in Burrendong Dam to save Dubbo from flooding. Wetlands usually rely on floods for their water (Williams 1981). When the river runs down effluent streams or spreads onto the floodplain, wetlands are filled. If there is less water, then fewer wetlands will fill and those that do fill will not last as long.

Large reservoirs and water diversions combined with increased ability to pump so-called 'surplus flows' into off-river storages have reduced flows. The amount of water in flows in the rivers of the

Murray-Darling Basin has declined. Average annual floods in the Murray River are 50% smaller than under natural conditions (Maheshwari *et al.* 1995). Severe droughts in the river that occurred in 5% of years before regulation now occur in more than 60% of years (Murray-Darling Basin Ministerial Council 1995). The algal bloom which affected 1000 km of the Darling River in 1991 was primarily due to high nutrient levels and low flow conditions, the latter exacerbated by extraction of water for irrigation (Bowling and Baker 1996). Species diversity in fish was negatively correlated with increased regulation of the river (Gehrke *et al.* 1995).

Floods which rejuvenate floodplains and provide water for other wetlands became less frequent and smaller in duration. This became most apparent in rivers whose water is expended in large terminal wetlands. Arguably the best known example of this is the Macquarie Marshes, although other examples exist (eg Gwydir, Lachlan, Namoi). The Macquarie Marshes well qualifies as a wetland for discussion because, as well as this significant problem, it suffers nearly all of the other wetland degradation problems which occur elsewhere in the Murray-Darling Basin.

Dams and weirs were also built over wetlands or wetlands were turned into storages for the water which was shunted to and from irrigation areas. Menindee Lakes, Lake Brewster and Lake Victoria are some of the more prominent examples. These wetlands were also the first to show the impacts of river management. Wetland vegetation adapted to wet and dry phases was drowned. Red Gums *Eucalyptus camaldulensis*, Black Box *E. microtheca* and Coolibah *E. coolibah* cannot withstand prolonged flooding. Similarly other aquatic vegetation such as Lignum *Muehlenbeckia florulenta* degrades and eventually dies if flooded for too long. Dead floodplain eucalypts usually scar the edges of these wetlands as a reminder of the damage caused by permanent flooding. The Murrumbidgee and Murray River wetlands are clear examples of the degradation caused by permanent flooding (Pressey 1990; Smith and Smith 1990; Briggs *et al.* 1994). Past vegetation covered Menindee Lake but it is now devoid of any vegetation (Kingsford 1995a).

The Macquarie Marshes — a wetland in crisis

The Macquarie Marshes in north-western New South Wales (Figure 13.1) has been recognised for its importance for conservation since the beginning of the twentieth century (Kingsford and Thomas 1995). The first recorded European to find the Macquarie Marshes was Oxley who was turned back in 1818 by the extensive reed beds which certainly reflected a wetland that had recently flooded. His descriptions have become a strong focal point for conservationists concerned about the future of the Macquarie Marshes and were given as evidence that the Macquarie Marshes was severely degraded. Unfortunately for the cause of conservation groups, the observations of Sturt, just 10 years later, were radically different. He was there in a dry period and scarcely paused on his way to the Darling River. So the debate raged that the Macquarie Marshes was sometimes dry and sometimes wet. Its ecological importance was recognised as it was listed as a wetland of International Importance under the Ramsar Convention and part of the Macquarie Marshes became a nature reserve in 1971.

Following the building of Burrendong Dam in 1967 and other river management structures (Figure 13.1), a significant irrigation industry became established. This was supported by the building of other dams and river management structures to control water in the catchment. In 1993–94, about 543 000 megalitres (1 megalitre = 10^6 litres) of water were diverted for irrigation. The total median annual flow is 940 000 megalitres, measured at Narromine upstream of the Macquarie Marshes

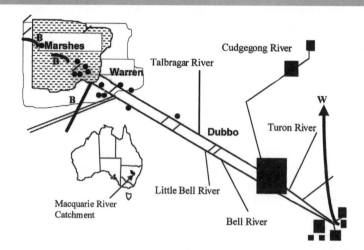

Figure13.1   Location of the Macquarie Marshes and schematic diagram of river management structures on the Macquarie River, showing inflowing rivers. Blocked-in rectangles represent nine major dams. Lines across the Macquarie River represent eight weirs. Four bypass channels are represented by B; filled in circles represent 12 regulators, cuttings, siphons, block dams and groynes; and W represents a water transfer scheme.

(Environment Protection Authority 1997) and so in 1993–94 about 58% of this median flow was diverted. The impact on the Macquarie Marshes was significant. They became at least 40–50% smaller than they used to be (Kingsford and Thomas 1995). Numbers of waterbirds and numbers of waterbird species have declined significantly over a 13-year period (1983–95). Many other aspects of wetland flooding were affected (see Kingsford and Thomas 1995). Critically, water quantity was shown to be irrefutably linked to the extent of wetland flooding. This was an extremely important point because it allowed time series analyses of 50 years of hydrological data that showed long-term degradation of the wetland. Importantly for the debate that followed, the implication was that unless something was done about the increasing diversion of water in the catchment, the Macquarie Marshes would continue to disappear.

The 1996 Macquarie Marshes Water Management Plan

In August 1996, the NSW Government released its 1996 water management plan for the Macquarie Marshes (Department of Land and Water Conservation and National Parks and Wildlife Service 1996) and it was implemented almost immediately. This followed an extensive discussion about the merits of the plan and submissions to the government. The crux of the debate was the focus of the Macquarie Marshes Water Management Plan (MMWMP). The plan did not simply concentrate on water management within the Macquarie Marshes but dealt with water management in the whole of the Macquarie River (see below). The implications were significant because it affected policies on diversion of water upstream and reliability of supply for the irrigation industry which was allocated more than 88% of water extracted from the Macquarie River; industry and towns take most of the rest (Department of Water Resources 1991).

The MMWMP aimed '… to identify and secure flows, from a finite water resource shared with others, to ensure the ecological sustainability of the Macquarie Marshes' (Department of Land and Water Conservation and National Parks and Wildlife Service 1996). Eleven rules govern access to water. The main ones were an increased allocation of water to the Macquarie Marshes of 75 000 ML which was in addition to 50 000 ML already provided and a limit on the extraction of off-allocation water to 50 000 ML. Off-allocation refers to flows not controlled by the dam which are additional to an irrigator's allocation. High flow capacity pumps are installed on the water course to pump these flows into large off-river storages. The timing of off-allocation access is determined by the extent to which a dam spills or tributaries downstream of the dam contribute their water to that in the river. In 1991–92, the amount of this water peaked when 200 000 ML of off-allocation water was diverted from the Macquarie River for irrigation.

The genesis of the Macquarie Marshes Water Management Plan followed a vocal and difficult debate within the community and government. Accusations were levelled that there was little community support for the plan (see Brock 1996), which were clearly wrong (see Tables 13.1–13.4). The main issues around the debate were articulated by the parts of the community most likely to benefit or suffer from changes to policies. The irrigation community generally did not support the plan. 'The Government by this Marsh Plan may be sending wrong signals to NSW Farmers and the community in general. In particular, it is by implication criticising the action of farming by irrigation when it is the means to addressing and delivering the world food needs' (Table 13.1). In contrast, the grazing community believed the plan represented a significant recognition of their rights to water. Previously these had been eroded as water was diverted upstream. The Macquarie Marshes Water Management Plan was '… a step in the right direction — we've been getting a pretty rough time in recent years' (Table 13.2). The Macquarie Marshes Environmental Landholders Assocation was established to represent graziers on the floodplains of the Macquarie Marshes. A bumper sticker articulated the importance of linking grazing with environmental sustainability: 'fat ducks mean fat cattle'. As well as the environmental costs of diverting water from the Macquarie Marshes, there have been significant impacts on the human community as people have left the district (McHugh 1996).

Significant public debate about the future of the Macquarie Marshes occurred between 1994 and 1996, the revision period of the MMWMP. Comments about the plan and different points of view about water management provide an insight into the critical issues for industry, government and the community.

Irrigation industry

Cereals, citrus, grapes, vegetables and cotton are irrigated in the Macquarie River catchment. About 93 000 ha is irrigated (Morrison and Bennett 1997). The actual amount diverted each year is dependent on the amount of water stored in Burrendong Dam and rainfall. In 1993–94, the amount of water diverted for irrigation peaked at 543 000 ML, following an increasing trend (Kingsford & Thomas 1995). Cotton accounts for over 50% of all water used for irrigation in the valley (Department of Land and Water Conservation, unpublished data).

The irrigation industry was highly critical of policy changes proposed in the MMWMP. In the rural town of Trangie, 1200 people demonstrated against the policy. The 1600 submissions to the draft plan were dominated by irrigation concerns, although many of these were 'form' letters. A

Table 13.1 Published responses (see source) from the irrigation industry to the problem and solution of the Macquarie Marshes. Direct quotes wee used as reported in the media.

Arguments	Comments	Source
Economic	Keep in mind that irrigation farming is central to the survival of people on this planet, that you produce almost a third of the production of this valley from one percent of the land.	Land 14/9/95, p. 114
	... agriculture feeds and clothes humans as well as caring for birds and animals and the environment ...	Land 5/10/95, p. 10
	Without irrigation you wouldn't have any cotton industry at all. In the last four years cotton has become increasingly significant because of the economic returns from cotton and that does supply a lot of employment in the Macquarie Valley.	
	If there was money in wetlands ... we'd all be in wetlands	McHugh (1996) p. 87
	There's only so much water in the bucket and that water has a value	McHugh (1996) p. 87
	We are talking well over $100 million in lost revenue to the valley ... The immediate loss of 750 jobs will have a devastating effect on the region.	Australian 15/9/95, p. 13
	The Government by this Marsh Plan may be sending wrong signals to NSW Farmers and the community in general. In particular, it is by implication criticising the action of farming by irrigation when it is the means to addressing and delivering the world food needs.	Media Associates (1997)
	And indeed it is the few graziers who stand to benefit from the current plan who are the Government's loudest proponents of this scheme rather than the majority in the valley who are going to be severely affected.	Media Associates (1997)
	For every 1000 megalitres of water that is used in the irrigation industry basically creates a job. These changes have the potential to lose somewhere between 750 and 1200 jobs for the Macquarie Valley.	Media Associates (1997)
	Cotton growers have had the most potential to convert a megalitre of water into dollars compared to most other things we are doing on farm.	Media Associates (1997)
	Allocating more irrigation water to the Macquarie Marshes at the expense of farmers would be like tipping more water into a leaking bucket ...	Land 31/8/95, p. 10
	We believe that water that goes to the Marshes is wasteful and that we can use water to better advantage than they can in the Marsh country.	ABC Quantum 27/6/961
Efficiency/ Management	Remember the marshes are Australia's largest irrigation system. If an irrigation property was operated in a similar manner, I doubt the owner would be in business for very long.	Land 12/10/95, p. 10
	I am quite convinced that cotton is not endangering the marshes. I think water management is. If an irrigation farmer ran their farm the way that the marshes are run at the moment, the irrigation farmer would go broke.	McHugh (1996) p. 87
	... the wetlands had received the 'lion's share' of the available water supply over the past eight years, but through lack of proper management it had not been utilised properly.	Land 31/8/95, p. 10
	The Macquarie Marshes do not need any more water, but they certainly need better management of the approximately 400,000 megalitres which they have received on average over the past 10 years	Land 28/9/95, p. 11
	Artificially flooding the marshes kills a lot of the native ground-laying bird wildlife. Too much water will also kill the native trees; their survival hinges on some dry times, too. There is sufficient water for the tree nesting variety.	Land 5/10/95, p. 10

Unfortunately, you can't throw water at the problem. Over the years, the creeks into the marshes have been channelled so that small flows down the river just keep on flowing through the area without giving one ounce of benefit.

Land 12/10/95, p. 10

The plan proposes to take a chunk of water and throw it at the marsh and that then becomes the solution to the marsh problem. Our plan is aimed at better utilisation of water so that maybe the same big volumes aren't necessary, but the marshes still are the number one priority.

Sydney Morning Herald 26/9/95; p. 3

Sending more water to the Marshes without improved management will not assist anything ... the Department of Conservation and National Parks and Wildlife Service are not capable of providing the required management.

Land and Water
Land 8/8/96, p. 3

... new laws will send excessive water to the marshes, forcing water tables to rise and exacerbating salt pollution in the already fragile bird-breeding grounds.

Sydney Morning Herald 13/12/95; p. 6

... continuous low-level flows provided since the dam opened had failed to mimic the natural wet and dry cycle of the wetlands and this was causing salt pollution.

Sydney Morning Herald 26/9/95; p. 3

It's a waste if it doesn't get sufficient overbank flooding to enhance the vegetation types. It's a waste if it doesn't cover enough of the grazier's country to give them the right amount of winter feed.

Media Associates (1997)

One other area that's contributed to degradation in the Marshes has been the constant small flows that go down the system throughout the year.

Media Associates (1997)

To maximise the efficient use of water on the marshes, capital investment is required in structures and banks.

Land 12/10/95, p. 10

I don't think irrigators are unhappy with the fact that water is going to the marsh. What irrigator's major concern is that there are other forms of degradation that have been identified in the Marsh that are not being addressed.

ABC Quantum 27/6/96[1]

... part of the marshes are now an artificially permanent wetland with associated problems of erosion, rising water table, salinity and dead trees.

Land 12/10/95, p. 10

Taking water away from the irrigation industry in the Macquarie to give to the marshes, while ignoring other complex management issues such as the impact of erosion and changes to flow patterns is neither good water nor good environment management.

Land 12/10/95, p. 10

The managers [of the Marshes] should be doing more with the water that they are currently getting.

Media Associates (1997)

The real question is whether the water now available for the marshes is being managed effectively by those responsible for them.

Land 25/5/95, p. 7

... proper examination of how en route storage or the building of another dam would help this process must be pursued.

Land 9/1/97, p. 24

Ecological values

If the flooding's come at the right time, like about spring, the birdlife then is absolutely amazing ... So we don't think we're interfering with them too much ... its certainly rich in birdlife and I would fight to see it maintained that way.

McHugh (1996) p. 81

... if one ibis looks like dying they'll send down a million gallons of water! Plus they've got 50,000 megs of water, which is roughly ten million dollars of water every year. So it should be used productively ...

McHugh (1996) p. 78

There's no use supporting two million birds if we only need 20 000.

McHugh (1996) p. 84

[1] Newsreel footage when the Burrendong dam was opened in the 1960s

Table 13.2   Published responses in the media (see source) from the grazing industry to the problem and solution of the Macquarie Marshes.

Arguments	Comments	Source
Impacts	Need a bridge because there was water under it for nine years. Now you wouldn't think so. We can go a full twelve months without getting water underneath it ... It's just gone to the irrigation industry–it's a dramatic increase in irrigation over recent years.	ABC *Quantum* 27/6/96
	You've got a wetland that used to go dry every ten years — now it's dry nearly all the time.	McHugh (1996) p. 75
	When Burrendong was built everyone noticed the drought was longer than before and we had no water for three years while it was filling.	*Sun-Herald* 9/10/94, p. 101
	A lot of the properties amalgamated because hundreds of hectares that once flooded are now dry and you can't do anything with it. It won't respond to rain.	*Sun-Herald* 9/10/94, p. 101
	My property, and several neighboring properties, used to receive natural flooding caused by rain in the upper catchment of the Macquarie, but since the virtually unchecked development of the Macquarie it is now nothing more than a trickle.	*Land* 1/12/94
	I have had to reduce my stock numbers by a third. You belt in and keep going but there are going to be a lot of people who won't be able to sustain and justify it and will have to get out.	Media Associates (1997)
	Historic pattern was that we got at least three floods a year. Our country was wet for up to 6–8 months. But for the last 15 years at least we've been cut back from probably a year like that every nine out of ten to three out of ten. We have lost population in the Marshes and also our ability to earn an income has been severely curtailed.	Media Associates (1997)
Water distribution	... and it seems a little ironic that irrigators in the Macquarie Valley and on off-river schemes can receive up to 80 per cent of their water allocation for growing crops such as cotton, and yet the lower Macquarie can be spared so little water that the river has virtually stopped flowing.	*Land* 1/12/94
	Irrigators ... all reckon they own the water.	McHugh (1996) p. 83
	Water used in the marshes, we used to say it wasn't a loss-it wasn't a loss, its the use of water, not a waste.	McHugh (1996) p. 88
	Under the natural conditions, your water rises and falls and your stock feed back as the water falls, then they come back again, but with releases, it stays the same for so long, it just keeps creeping up and you end up with no country to run sheep on.	McHugh 1996, p. 84
	... a step in the right direction — we've been getting a pretty rough time in recent years [describing the Macquarie Marshes Water Management Plan].	*Land* 14/9/95, p. 114
Ecological values	I haven't ever seen it [bird rookery] as dry as it is now out the back — ever.	Media Associates (1997)
	In a wet season its well I suppose to put it in a word it would be a paradise as far as people who love and appreciate birds.	Media Associates (1997)
	In the last 10 years we've had two breeding programs with our waterbirds out of ten ... Pre that we were probably getting six to eight out of ten.	Media Associates (1997)
	I think our natural ecology has got to be looked after and managed and we haven't been doing that.	Media Associates (1997)
	It will never go back to pre Burrendong which was in 1966. But we hope to restore some of the Marshes to their natural beauty that they were and by using best management practices I think the grazing industry now is so much more aware of what can be done and what will be done that we hope to be able to plan our future a lot better than we did before.	Media Associates (1997)

Table 13.3   Published responses in the media from government to the problem and solution of the Macquarie Marshes.

Group	Comments	Source
Conservation	In general, water is over-allocated over the whole of the Murray-Darling Basin. In some dry periods, we know that water simply doesn't reach the sea. We use it all and that isn't sustainable.	Media Associates (1997)
	We're very keen to use the Convention to encourage the New South Wales and Commonwealth Governments to manage the whole Marsh ecosystem, the catchment, the entire Marsh as well as the 10% of land that is actually in the Nature Reserve and listed under the Ramsar Convention.	Media Associates (1997)
	The irrigation community needs to understand that water is a limited resource, and plan accordingly.	Land, 8/8/96, p. 3
Unknown	Before cotton was grown in the Macquarie Valley the marshes were a delightful place to visit. Water birds of all types would be seen, but that has virtually disappeared due to lack of water — another part of heritage lost.	Land, 28/9/95, p. 11
	… the loss of water that the irrigators are angry about is off-allocated water — not their licensed (allocated) water. More (not all) of this 'free flow' non-allocated water can come down to the wetlands and sheep and cattle pasture...	Land, 28/9/95, p. 11
	We are delighted that we have a Government that is prepared to bite the bullet on this difficult issue. We have been pressing successive governments to address the issue of over-allocation and over-extraction of water and to restore the health of the rivers.	

Table 13.4   Published responses in the media from government to the problem and solution of the Macquarie Marshes.

Group	Comments	Source
New South Wales National Parks and Wildlife Service	The lack of water is the biggest threat but one of the most critical ways of addressing that is to manage the limited amount we have to maximum efficiency — and the first thing is to manage erosion.	McHugh (1996) p. 76
	When you take flooding off this wet country [Marshes] which used to be couch and all the waterplants … You're having no dryland species in the soil. All you get are the invasive species like the rolly-polly and the buckbush and those thistles … But as far as productivity from the Nature Reserve point of view as feeding area for the birds, we've lost all that.	Media Associates (1997)
	Cotton-growers do understand the value of water. With our Wildlife Allocation…they might not believe in it, they might not think its being used the right way, but they see that if we have a wildlife allocation, then we don't want it running away in channels, we want it spreading out across the marsh.	McHugh (1996) p. 86
	They say, "We're producing a crop worth X million dollars, what are you doing?" Well, we're producing X thousand waterbirds.	McHugh (1996) p. 84
	We've got this area of trees which are getting quite stressed from water and many are dying	Media Associates (1997)
	I personally feel that it is very hard in my position with National Parks to have to go to meetings and talk to people and actually defend that we are trying to get water for the environment. I personally think that it should be the other way around.	Media Associates (1997)
	We're not going to get the South Marsh back to even 50% flood but what we have again is a big gain and the wildlife particularly has gained.	Media Associates (1997)
	The low flows, the trickles that are causing in combination with say carp and other things, erosion which is causing a lot of problem and draining the Marsh. It's all right to 'um' and 'ah' and carry on. We did that for 25 years … and nothing was done and that's why we have got these big erosion channels.	Media Associates (1997)
	What we realised as we did our aerial surveys was that there was less and less waterbirds here and the area that was being flooded was actually decreasing and it was about time we had a good hard long look at this. So we matched the waterbird data that we had with some of the hydrology…	ABC Quantum 27/6/96
	So what we're trying to do here is say let's have a certain amount of water that's guaranteed for the environment and then let's see if we can keep a wetland that is functional. It's still not going to function the way it did early this century. But we still have a very exciting place.	ABC Quantum 27/6/96
	It is such a complex system. We've got a certain amount of water. How are we best going to utilise it? Do we have a big flood once every five years or do we strive for a median flood once every three years? They are very interesting problems. And I think those are the challenges, not just for National Parks but for the whole community.	ABC Quantum 27/6/96
	Look everybody knows that feral animals are a problem and weeds are a problem and erosion is a problem in these sorts of systems and similarly grazing may be a problem in some cases. But the really big problem here is lack of water.	ABC Quantum 27/6/96
	Regulation and extractive water use have changed the nature of the flows in the Macquarie as surely as the climate had changed.	McHugh (1996) p. 76
Department of Land and Water Conservation[a]	… everybody showed concern for the marshes. We need to try and fix it up,be smart with what water we do have to stop it draining — that was the general feeling.	McHugh (1996) p. 86

Department of Land and Water Conservation[a]	They were unauthorised banks … not illegal works. We're not land managers, we're water managers	McHugh (1996) p. 77
	The impact of development, the dam and the irrigation industry has been to reduce the flows in the middle range, the median sized floods and so the Marsh is left basically with a sequence of dry years or the big floods.	Media Associates (1997)
	The analysis of the flows over the long-term shows that the flow to Marsh has been quite variable from year to year and that variably in flow has created the diversity of habitat that we have in the Marsh and so the diversity in the Marsh is going to be impacted unless we reinstate those median size floods.	Media Associates (1997)
	The task for the environmentalist is to prove the value of protecting the marshes in comparison to export income derived from irrigation.	McHugh (1996) p. 87
Local Council	Licences in my opinion has been totally over granted. There have been licences granted to a lot of landholders that have remained unused … The Water Schemes that were allowed to develop away off the Macquarie were too far away from the river. In one scheme alone – there is 360 km of channels. Now that means that water goes up those channels. There is evaporation, there is seepage, cost of maintaining those channels.	Media Associates (1997)
	Everybody wants the Marshes to remain because they're a part of Australia's history and they're a very big part of the environment.	Media Associates (1997)
	Just giving more water at times is not the answer to the problem. If we put more water down to the Marshes, you're taking water off people who are creating wealth in this region. Creating wealth is part of ongoing development of Australia.	Media Associates (1997)

[a] Formerly the Department of Water Resources

cabinet meeting of the NSW Government in Dubbo was also disrupted by protestors. There were two main criticisms of the policy changes proposed in the MMWMP: economics, and efficiency and management of water within the Macquarie Marshes. The main economic argument was that a reduction in allocations of water to the irrigation industry would cost jobs and affect the economy of the region. 'We are taking well over $100 million in lost revenue to the valley…The immediate loss of 750 jobs will have a devastating effect on the region' (Table 13.1). As well, argument was mounted that the real reason for the degradation of the Macquarie Marshes was not because of upstream diversions of water but inefficiency and mismanagement of water within the Marshes. 'Taking water away from the irrigation industry in the Macquarie to give to the marshes, while ignoring other complex management issues such as impact of erosion and changes to flow patterns is neither good water nor good environment management' (Table 13.1).

Economics

Economic arguments from the irrigation industry centred on the pivotal role of irrigation for the welfare of human communities in general and, in particular, the importance of the irrigation industry to the economic health of the Macquarie Valley. 'Keep in mind that irrigation farming is central to the survival of people on this planet, that you produce almost a third of the production of this valley from one percent of the land' (Table 13.1). But the natural resource in short supply is not land but water. In the Macquarie Valley, irrigation had entitlements to more than 88% of the water (Department of Water Resources 1991).

The primary issue was access to water. The concern for the irrigation industry was that the new policies (see above), which cut back diversion of water by 12%, constrained irrigation and hence economic output in the Macquarie River Valley. For example, 'Cotton growers have had the most potential to convert a megalitre of water into dollars compared to most other things we are doing on farm' (Table 13.1). Irrigated agriculture in the valley had expanded from 17 500 ha in 1965 to 85 577 ha in 1990 and 93 000 ha in 1993 (Kingsford and Thomas 1995; Morrison and Bennett 1997). The trend in diversions of water mirrored this expansion. The logical development of an argument predicated on unrestricted access to water would be to sanction increased diversion of water. This would certainly lead to increased economic returns to the valley and probably the State of New South Wales in the short term. For example, irrigated cotton in the Macquarie River Valley produced $100 million in 1993 (Department of Land and Water Conservation, unpublished data).

The question would still remain about what limit there should be on water diversions from the Macquarie River, given that it is not a limitless resource. The Macquarie Marshes are dependent on water provided from upstream and degradation will increase with increasing diversions and possibly even if held at levels before the MMWMP. There may even be a lag effect, with degradation continuing after flows are restored by the plan. Sentiments that water might be wasted in terms of irrigated agriculture — 'We believe that water that goes to the Marshes is wasteful and that we can use water to better advantage than they can in the Marsh country' (see Table 13.1) — ignores the fact that it sustains complex freshwater ecosystems. If diversions were allowed to continue, eventually, like the most notorious example of degradation of a freshwater ecosystem (the Aral Sea in Uzebekistan; Micklin 1988), the Macquarie Marshes could collapse entirely. And what of the significant environmental degradation which has already occurred? 'Allocating more irrigation water to the Macquarie Marshes at the expense of farmers would be like tipping more water into a leaking bucket …' (Table 13.1). This statement ignores the fact that the 'irrigation water' used to be 'Marshes water'. To restore the Macquarie Marshes to its original state is impossible and socially unacceptable because it would mean removing the dams and preventing access to water but there was a recognition of the ecological cost already paid as policy makers searched for the compromise.

Is the economics as clear cut as some proponents of the irrigation industry argue? 'There's only so much water in the bucket and that water has a value' (Table 13.1). Financial subsidies for water in Australia in 1994 were estimated to amount to $3.3 billion which included no estimate of environmental subsidies (Department of the Environment, Sport and Territories 1996). The subsidy for rural water in NSW was estimated to be $400 million in 1992–93 with total revenues only meeting 25% of total expenses (Department of the Environment, Sport and Territories 1996). Wetlands and freshwater ecosystems also have an economic value (Morrison and Kingsford 1997). For example, increased flooding leads to higher production for graziers (see below). Water quality, which is related to flows, affects fishing and recreation. Wetland areas such as the Macquarie Marshes may have significant tourism potential. Already, at least one landholder in the Macquarie Marshes offers accommodation for visitors (Cunningham 1997). Open days at the Macquarie Marshes Nature Reserve are well attended. Certainly wetland areas in inland Australia are attracting increasing numbers of visitors (Kingsford and Halse in press). There may also be significant intrinsic economic value in some wetlands, particularly icons such as the Macquarie Marshes. All economic values are theoretically measurable but seldom reach the policy table because of the difficulties of measurement. Thus estimates of the impact of restricting diversions of water on the irrigation community of 6% (Department of Land and Water Conservation and National Parks and Wildlife Service 1996) took no account of the positive economic impact of flows, even for more easily measurable variables such as livestock production as a result of grazing benefits. As well, significant economic costs may occur if current trends of increasing salinity in the Macquarie River continue (Williamson *et al.* 1997). Repair of freshwater ecosystems can incur a significant cost. Arguments which translate loss of water to irrigation into direct economic impact and job loss are simplistic. Certainly there are likely to be financial impacts on individuals which may depend on their ability to absorb the costs.

The economic challenge for the irrigation community is to sustain growth using the same amount of water. Significant cost savings could be made in water use (see Postel 1992). These savings may be registered in terms of increasing efficiency, such as drip irrigation. The challenge will increasingly be to do more with the water which is available.

Efficiency and management of water

The most serious criticism levelled at the new government plan (MMWMP) was that it had ignored the most significant issue: management of water within the Macquarie Marshes. Essentially the irrigation industry maintained that '[t]he real question is whether the water now available for the marshes is being managed effectively by those responsible for them' (Table 13.1). One definition of efficiency was linked to efficiency of running an irrigation enterprise. 'Remember the marshes are Australia's largest irrigation system. If an irrigation property was operated in a similar manner, I doubt the owner would be in business for long' (Table 13.1). The difference between an agricultural enterprise and an ecosystem could not be greater. Irrigation manages water for one species in a landscaped and strictly controlled environment. Unlike agricultural enterprises, the wetland ecosystem is significantly more complex. Life cycles of thousands of species and at least hundreds of thousands of organisms interact in ways in which science is only scratching the surface in understanding.

As well as the efficiency argument, particular management issues were raised by the irrigation industry. These included prolonged flooding resulting in dead trees and invasion of Cumbungi *Typha* sp., channelisation, increased salinity, continuous low flows and the artificial flow regime. 'One other area that's contributed to degradation in the Marshes has been the constant small flows that go down the system throughout the year' (Table 13.1). Nearly all other river systems in the Murray-Darling Basin are plagued by similar problems, plus a few more (Walker 1985; Barmuta *et al.* 1992; Kingsford 1995a). They are symptoms of river regulation. Other issues include sedimentation, changed seasonality and temperature depression as a result of coldwater releases from the bottom of dams. Certainly the Macquarie River has problems with temperature depression. Temperature of water more than 300 km from the dam is 3–5°C degrees lower than it would be under natural conditions (J. Harris, personal communication). This can affect breeding of fish species.

What of the major management issues in the Macquarie Marshes nominated by the irrigation industry? They all exist and are all significant. Erosion of channels for example has caused significant impacts on the extent of flooding. The 1966 and 1967 floods were underestimated when modelled on recent data because of the impact of channelisation (Bell *et al.* 1983). Channelisation and continuous low flows cause erosion which then means a larger flood is needed for overbank flows which sustain the Marshes ecosystem. Erosion is most serious in the southern Marshes, on private land and in the nature reserve. There has been an increase in the size of some channels by 30%, since the late 1960s (Brereton 1994). Agencies and landholders are addressing this problem with erosion control works. It is a difficult issue to resolve because of local water-sharing debates and because of the way flows are managed in the river. Erosion is primarily exacerbated by flow management. Low flows are more frequent because this is how water is delivered to irrigators, within and downstream of the Marshes, and for stock and domestic supply. Too much of this sort of water flow leads to prolonged flooding of areas of the Macquarie Marshes and so trees die and vegetation adapted to prolonged flooding, such as Cumbungi, becomes established. These are all symptoms of poor water management for wildlife but good management for human use. Water is delivered as efficiently as possible for human use. The same arguments can easily be made about establishment of dams. They regulate and control rivers, allowing security of supply for irrigation and urban usage. They reflect good water management for human use but poor management for freshwater ecosystems. To argue that the major ecological problems of the Macquarie Marshes are due to poor government management is certainly true but it is this same management which has developed a management system which is geared primarily to efficient delivery of water to users, primarily for irrigation.

Concerns about the draft MMWMP prompted the Macquarie River Advisory Council (MRAC) to produce an alternative view of water management for the Macquarie River. MRAC was a group set up by the water manager (previously the Department of Water Resources) to advise on water management in the Macquarie River. Nine of the 12 members on the committee represented irrigators or had irrigation interests and there were no environmental representatives in late 1995 (Media Associates 1997). There were five general principles articulated in the MRAC report. The first was to mimic wetting and drying cycles. Next, there was an objective to identify the minimum amount of water necessary for ecological sustainability of the Macquarie Marshes. This was followed by an objective to specify ecological outcomes of the 1996 MMWMP in the Macquarie Marshes. The final principle was related to water access: to establish guidelines for the provision of the maximum amount of water possible for irrigators in the Macquarie Valley.

There is little ecological argument about the necessity to mimic wetting and drying cycles in the management of wetlands and rivers in Australia although the practicalities of doing so are limited. Large dams and commitments to providing water for irrigation and towns necessarily mean that our regulated rivers will never truly mimic natural wetting and drying cycles. The next two principles are particularly difficult to translate into practical outcomes. Due to the complexity of aquatic ecosystems, there are difficult questions about lack of data, the definition of ecological sustainability, the identification of ecological outcomes and specifying the minimum amount of water. Finally specification of the 'maximum amount of water possible for irrigators' is problematic as I have argued above in terms of ecological sustainability.

Grazing industry

Most of the Macquarie Marshes is owned or leased by private landholders. Only about 14% of the Macquarie Marshes is designated as a nature reserve managed by NSW National Parks and Wildlife Service (Kingsford and Thomas 1995). Nearly all of the landholders rely on floods within the Macquarie Marshes to support their livelihoods. Production from grazing in the Macquarie Marshes, measured as regional gross margin for 1996, was estimated to be $5.2–7.5 million with wetland areas supporting considerably more stock than dryland areas (Cunningham 1997). Of 27 landholders surveyed within the Macquarie Marshes, all believed that river regulation and water diversions had had significant impacts on the area (Cunningham 1997). The science (Kingsford and Thomas 1995) supported their anecdotal observations of the state of their land and ecological values within the Macquarie Marshes. 'Historic pattern was that we got at least three floods a year. Our country was wet for up to 6–8 months. But for the last 15 years at least we've been cut back from probably a year like that every nine out of ten to three out of ten. We have lost population in the Marshes and also our ability to earn an income has been severely curtailed' (Table 13.2).

Their support for the MMWMP 1996 was strong and their comments on the new policy changes reflected this. 'It will never go back to pre-Burrendong which was in 1966. But we hope to restore some of the Marshes to their natural beauty and by using best management practices I think the grazing industry now is so much more aware of what can be done and what will be done that we hope to be able to plan our future a lot better than we did before' (Table 13.2). Of 112 landholders with holdings that included riparian ecosystems between the southern end of the Marshes and the Barwon River (mainly graziers), 109 (97%) supported the initiative to restore flows to the Macquarie Marshes through the policy changes put forward in the Macquarie Marshes Water Management Plan. Unlike many other resource management issues, the role of conservation groups in the Macquarie Marshes was relatively minor compared with that of the riparian grazing community within the Macquarie Marshes. Without the support of the grazing community, the major policy changes may not have enventuated. As remarked by someone in the irrigation industry: 'And indeed it is the few graziers who stand to benefit from the current plan who are the Government's loudest proponents of this scheme rather than the majority in the valley who are going to be severely affected'. It will be a challenge to see if graziers consider impacts of grazing on wetlands (Robertson 1997) and then manage the wetland ecosystems with the same commitment made to sustainable water management.

Conservation groups

Relative to other major environmental issues (eg forestry, greenhouse), conservation groups have not become significantly involved in the state of Australia's rivers (Kingsford and Halse in press). This may be because the issues are remote and the science not comprehensive (Kingsford 1995b;

Kingsford and Halse in press). Also, water management is possibly a more complex issue to understand. Certainly, the emphasis of their comments was supportive of government initiatives to restore some of the flows to the Macquarie Marshes. 'The irrigation community needs to understand that water is a limited resource, and plan accordingly' (Table 13.3). The relative paucity of comment relative to the irrigation and grazing industry reflected their remoteness from the issue and conversely the importance of the issue in the daily lives of irrigators and graziers.

Government

Scientists and managers in government also provided comment and expertise on the issue. They may also represent the wider communities' interests. For example, the National Parks and Wildlife Service should represent conservation interests, as these are reflected in the *National Parks and Wildlife Act 1974*. There was general agreement that the flows to the Macquarie Marshes had decreased and affected the fauna and flora dependent on water (Table 13.4). Generally, comments by government employees about the Macquarie Marshes reflected the problems with sustainability of the Macquarie Marshes. 'Regulation and extractive water use have changed the nature of the flows in the Macquarie as surely as the climate had changed' and 'The analysis of the flows over the long-term shows the flow to Marsh has been quite variable from year to year and that variability in flow has created the diversity of habitat that we have in the Marsh and so the diversity in the Marsh is going to be impacted unless we reinstate those median size floods' (Table 13.4). The difficulty of local government is expressed in their statements. There is an element of wishing to have the best of both worlds. 'Everybody wants the Marshes to remain because they're a part of Australia's history and they're a very big part of the environment' and 'Just giving more water at times is not the answer to the problem. If we put more water down to the Marshes, you're taking water off the people who are creating wealth in this region. Creating wealth is part of the ongoing development of Australia '(Table 13.4). Such an attitude is particularly prevalent in the general community in relation to conservation issues. Hence the importance of showing that such natural resource issues involve trade-offs.

Conclusion

Humans need water. We have controlled and manipulated it to suit our ends wherever possible. The maxim of Parakrama Bahu the Great, the twelfth-century king of modern Sri Lanka, has governed use of water everywhere in the world: 'Let not even a small quantity of water obtained by rain go to the sea, without benefitting man' (Postel 1992). The management of water in the Macquarie River reflected this philosophy until the release of the MMWMP. Except, like many of Australia's rivers where water does not reach the sea, one could easily substitute 'sea' with 'wetland'.

Humans have benefited enormously from development of water resources. Industries and towns have become established and prospered where none existed before. But this has come at a high environmental cost. Nowhere is that cost so well illustrated as in the Macquarie Marshes where 50 years of increasing diversions of water from the Macquarie River have significantly reduced the size of the wetland and affected the flow regime of the river with largely unknown consequences. Some in the community may argue that such degradation should be tolerated for the economic wealth that comes from water resource development.

One of the most difficult lessons to learn from this issue is that diversion of water from rivers will incur an environmental cost on wetlands and this often translates into a later economic cost (increasing salinity, reduced water quality, loss of wetlands for tourism). Importantly, diversions of water can also divide rural communities. It is not just the Macquarie where the rural community is divided over water. Landholders on the Gwydir River, Narran River, Paroo River and Cooper Creek are divided into different camps. There are winners and losers in this exploitation in the short term. This is no different to other parts of the world where the significance of water can even force nations to consider war (Postel 1992). There is now sufficient information to have an informed and vigorous debate about the long-term sustainability of rivers and wetlands and the costs of diverting water before sanctioning further diversion of water.

Reassessment of the primacy of a water exploitation culture followed by realignment of policies which recognise the ecological impacts of this exploitation necessarily mean that communities whose livelihoods are built around access to water will be affected. This is what happened in the Macquarie Marshes. Release and implementation of the MMWMP reduced the allocations to the irrigation industry and allowed more water to flow to the Macquarie Marshes and the floodplain graziers. Some may have capitalised infrastructure in the expectation that access to water would continue to be allowed. Successive governments since the 1950s have, until recently, promoted the exploitation of water resources. Decision makers need to understand the socio-economic impacts of such changes to some in the rural community. That's why reversal of ecological degradation on most river systems will be difficult in the short term. Changes of the order of those needed in the Macquarie Marshes are not easily achieved. This difficulty raises questions about river systems where little water is diverted. There is significant economic pressure to extract water from pristine river systems (eg Warrego River, Paroo River, Bulloo River, Cooper Creek and Fitzroy River).

Test cases such as the Macquarie Marshes highlight the difficult questions which rural communities and governments need to resolve with regard to sustainable water use. Do we divert more water from our rivers for increased economic output in the short-term and live with the environmental degradation that ensues or should we take stock of exploitation of our river systems? The issue that is critical to this argument is an acknowledgement that the water supply is not limitless and, importantly, extraction of water has a significant impact on freshwater ecosystems. This is the main reason why the four State governments which manage the Murray-Darling Basin agreed to a cap on diversions of water from the valleys (Murray-Darling Basin Ministerial Council 1995).

Major challenges face rural communities as access to water becomes limited, as we head toward 2000. Reducing exploitation of water resources will be particularly difficult to achieve despite government policies and initiatives to restrict access to water (eg the Murray-Darling Basin cap). There will always be rivers whose flows are not yet diverted which will be considered to have potential. Such a sentiment is expressed by the previous Queensland Government in its assessment of the Murray-Darling Basin Cap (Murray-Darling Basin Ministerial Council 1995). It argues for increased development of water resources in Queensland because development so far has been slow. Growing populations in Australia and worldwide will provide the momentum to push development harder. The jury will revisit this issue in the year 2000. If we have continued to exploit the flows in our rivers, the cost to the wetlands will continue to be significant. If Australia cannot resolve these major challenges to sustainability in the management of water then there is little hope for countries whose rivers move between countries.

Acknowledgements

I thank Bill Johnson and two anonymous referees for providing useful comments on this paper.

References

Allan, J.D., and Flecker, A.S. (1993). Biodiversity conservation in running waters. *Bioscience* **43**, 32–43.

Anderson, I. (1996). World's wetlands sucked dry. *New Scientist* **149**(2023), 9.

Australian Nature Conservation Agency (1996). *A directory of important wetlands in Australia* 2nd edn. (Australian Nature Conservation Agency: Canberra.)

Barmuta, L.A., Marchant, R., and Lake, P.S. (1992). Degradation of Australian streams and progress towards conservation and management in Victoria. In *River conservation and management.* (Eds P.J. Boon, P. Calow, and G.E. Petts). pp. 65–79. (John Wiley and Sons Lts: Chichester).

Bell, F. C., Coates, S. J., and Smith, G. (1983). Resolution of a conflict between water needs for irrigation and conservation? The Macquarie Marshes. In *Proceedings of the Water Resources Conference, 16 July 1983, Macquarie University, Sydney.* (Eds P. Crabb, D.C. Rich, and S.J. Riley.) pp. 60–70.

Blackman, J.G., Perry, T.W., Ford, G.I., Craven, S.A., Gardiner, S.J. and De Lai, R.J. (1996). Queensland. In *A directory of important wetlands in Australia* 2nd edn. (Eds R. Blackley, S. Usback, and K. Langford). pp. 177–433. (Australian Nature Conservation Agency: Canberra.)

Brereton, G.J. (1994). *An investigation into the impact of erosion in the Southern Macquarie Marshes.* Unpublished report to Macquarie Marshes Catchment Committee, Dubbo.

Briggs, S.V., Hodgson, P. F. and Ewin, P. (1994). Changes in populations of waterbirds on a wetland following water storage. *Wetlands (Australia)* **13**, 36–48.

Brock, P.M. (1996). Landholder views on the boundaries of the Macquarie Marshes. *Wetlands (Australia)* **15**, 72–91.

Bowling, L. and Baker, P.D. (1996). Major cyanobacterial bloom in the Barwon–Darling River, Australia, in 1991, and underlying limnological conditions. *Marine Freshwater Research* **47**, 643–657.

Cunningham, G. (1997). *Macquarie Marshes grazing study — 1996.* Prepared for the Macquarie Marshes Catchment Committee. p. 23.

Department of Land and Water Conservation and National Parks and Wildlife Service (1996). *Macquarie Marshes water management plan 1996.* p. 38. (National Parks and Wildlife Service and Department of Land and Water Conservation: Sydney.)

Department of the Environment, Sport and Territories (1996). *Subsidies to the use of natural resources.* (Department of the Environment, Sport and Territories: Canberra.)

Department of Water Resources (1991). *Water resources of the Macquarie Valley.* (Department of Water Resources, NSW: Sydney.)

Environment Protection Authority (1997). *Proposed interim environmental objectives for NSW waters.* (Environment Protection Authority: Chatswood, Sydney.)

Finlayson, C.M., and Moser, M. (1991). *Wetlands.* (International Waterfowl and Wetlands Research Bureau, Slimbridge, Facts on File Limited: Oxford.)

Gehrke, P.C., Brown, P., Schiller, C.B., Moffatt, D.B. and Bruce, A.M. (1995). River regulation and fish communities in the Murray-Darling River system, Australia. *Regulated Rivers: Research and Management* **11**, 363–375.

Goodrick, G. (1970). *A survey of wetlands of coastal New South Wales.* CSIRO Division of Wildlife Research Tech. Memo No. 5.

Halse, S.A. (1989). Wetlands of the Swan coastal plain past and present. In *Swan coastal groundwater management conference — proceedings.* (Ed G. Lowe). pp. 105–108. (Western Australian Water Resources Council Publication No. 1/89, Perth.)

Hollis, T. (1992). The causes of wetland loss and degradation in the Mediterranean. In *Managing Mediterranean wetlands and their birds.* (Eds C.M. Finlayson, G.E. Hollis, and T.J.Davis). pp. 83–90. (International Wetlands Research Bureau Special Publ. No. 20: Slimbridge, United Kingdom.)

Jones, D., Cocklin, C., and Cutting, M. (1995). Institutional and landowner perspectives on wetland management in New Zealand. *Journal of Environmental Management* **45**, 143–161.

Kingsford, R.T. (1995a). Ecological effects of river management in New South Wales. In *Conserving biodiversity: threats and solutions.* (Eds R. Bradstock, T.D. Auld, D.A. Keith, R.T. Kingsford, D. Lunney, and D. Sivertsen.) pp. 144–161. (Surrey Beatty and Sons: Sydney.)

Kingsford, R.T. (1995b). Occurrence of high concentrations of waterbirds in arid Australia. *Journal of Arid Environments* **29**, 421–425.

Kingsford, R.T., and Halse, S.A. (in press). Waterbirds as the 'flagship' for the conservation of arid zone wetlands of Australia? In *Wetlands for the future.* Proceedings of INTECOL'S V International Wetlands Conference, Perth 1996. (Eds A.J. McComb and J.A. Davis). (Gleneagles Press: Adelaide.)

Kingsford, R.T., and Thomas, R. F. (1995). The Macquarie Marshes in arid Australia and their waterbirds: a 50 year history of decline. *Environmental Management* **19**, 867–878.

Maheshwari, B.L., Walker, K.F. and Mcmahon, T.A. (1995). Effects of regulation on the flow regime of the River Murray, Australia. *Regulated Rivers — Research and Management* **10**, 15–38.

McComb, A.J. and Lake (1988). *The conservation of Australian wetlands*. (Surrey Beatty and Sons: Sydney.)

McHugh, S. (1996). *Cottoning on. Stories of Australian cotton-growing*. (Hale and Iremonger: Sydney.)

Media Associates (1997). *Where the River Runs dry*. Video. 42 minutes, Media Associates, Canberra.

Micklin, P.P. (1988). Dessication of the Aral Sea: a water management disaster in the Soviet Union. *Science* **241**, 1170–1176.

Morrison, M.D., and Bennett, J.W. (1997). *Water use trade-offs in the Macquarie and Gwydir valleys*. Research Report No. 2. p. 23. (University of New South Wales: Sydney.)

Morrison, M.D., and Kingsford, R.T. (1997). The management of inland wetlands and river flows and the importance of economic valuation in New South Wales. *Wetlands* (Australia) **16**, 83–98.

Murray-Darling Basin Ministerial Council (1995). *An audit of water use in the Murray-Darling Basin*. p. 40. (Murray-Darling Ministerial Council: Canberra.)

Norman, F.I., and Corrick, A.H. (1988). Wetlands in Victoria: a brief review. In *The conservation of Australian wetland*. (Eds A.J. McComb, and P.S. Lake). pp. 17–34. (Surrey Beatty and Sons: Sydney).

Postel, S. (1992). *Last oasis*. (Worldwatch Institute, W.W. Norton and Company: New York.) p. 239.

Pressey, R.L. (1989). Wetlands of the Lower Clarence Floodplain, Northern Coastal New South Wales. *Proc. Linn. Soc. N.S.W.* **111**, 143–155.

Pressey, R.L. (1990). Wetlands. In *The Murray*. (Eds N. Mackay, and D. Eastburn.) pp. 167–181. (Murray Darling Basin Commission: Canberra.)

Robertson, A.I. (1997). Land-water linkages in floodplain river systems: the influence of domestic stock. In *Frontiers in ecology: building the links*. (Eds N. Klomp, and I. Lunt). pp. 207–218. (Elsevier Science Ltd: Oxford, United Kingdom.)

Smith, P., and Smith, J. (1990). Floodplain vegetation. In *The Murray*. (Eds N. Mackay, and D. Eastburn.) pp. 215–228. (Murray Darling Basin Commission: Canberra.)

Walker, K.F. (1985). A review of the ecological effects of river regulation in Australia. *Hydrobiologia* **125**, 111–129.

Wasson, B., Banens, B., Davies, P., Maher, W., Robinson, S., Tait, D., and Watson-Browne, S. (1996). Inland waters. In *Australia: State of Environment 1996*. Chapter 7, pp 1–55. (CSIRO Publishing: Melbourne.)

Wiens, J.A., Patten, D.T., and Botkin, D.B. (1993). Assessing ecological impact assessment: lessons from Mono Lake, California. *Ecological Applications* **3**, 595–609.

Williams, W.D. (1981). Inland aquatic systems: an overview. In *Ecological biogeography of Australia*. (Ed. A. Keast.) pp. 1079–1099. (Dr W. Junk: The Hague-Boston-London).

Williamson, D.R., Gates, G.W. B., Robinson, G., Linke, G.K., Seker, M.P. and Evans, W.R. (1997). *Salt trends. Historic trend in salt concentrations and saltload of stream flow in the Murray-Darling Drainage Division*. Dryland Technical Report No. 1. p. 64. (Murray-Darling Ministerial Council: Canberra.)

Co-operative management of road reserves for biodiversity maintenance

Quentin Farmar-Bowers

Introduction

There is quite an extensive literature in Australia and overseas about the importance of linear reserve or corridors, for the conservation of biodiversity; for example see Saunders and Hobbs (1991), Breckwoldt *et al.* (1990) and Williams *et al.* (1997). There is also a growing general literature on the measures of biodiversity and its importance; see for instance Huston (1994) and Reaka-Kudla *et al.* (1997).

The conservation significance of road reserves lies to a substantial degree in that they are samples, albeit much changed samples, of the original ecosystems through which the road was built. The transport network in Australia is very extensive and probably crosses most of the major ecosystems on the continent. The information is not readily available yet, but a close inspection of maps and remote images of Australia leads to the conclusion that road reserves probably sample more of Australia's ecosystems than are protected in reserves. Transport corridors were not set aside as samples of ecosystems — it just happened.

The remaining strips of native ecosystems on road reserves are vulnerable to degradation from road construction, maintenance and use, and also from the adjacent land uses. It seems inevitable that the continuing application of current development technologies will eventually remove what survives of this grid of biodiversity. It is not planned but it may just happen. If somehow the biodiversity on transport corridors is maintained, then the country will retain a little more of its original biodiversity, and it will also retain a detailed living map of the distribution of this biodiversity.

Road reserves could be viewed as a kind of network of ecosystem arteries, perhaps now rather depleted of blood, but still in place and potentially able to expand and pump again with vitality. Road reserves are sometimes viewed as corridors for the movement or migrations of plants and animals as well as for humans. But as static samples of the original pattern of ecosystems, they also provide a corridor back in time and so provide an illustration of the original ecological processes occurring in the landscape. The connection between the ecosystems remaining on the road reserves and the surrounding country may provide an important guide as to 'why' as well as 'how' to maintain their values in perpetuity.

In areas where the land adjacent to the road still supports native ecosystems, the road reserve can function as a buffer protecting the adjacent ecosystems from change. Either way, the road reserves are playing an important role in maintaining Australia's biodiversity.

The developments and changes that are occurring daily are not purposely focused on destroying biodiversity; the destruction is occurring inadvertently. Action is needed on every road reserve to maintain the existing biodiversity and this implies that changes are needed in the planning and management processes employed today. The task appears enormous and the expertise and resources needed to identify and maintain biodiversity are not immediately available. In addition, time is ticking away. Australia has knowledgeable and skilful people who could find ways to maintain the biodiversity and concomitantly modernising road transport. Perhaps this can be achieved by setting dual objectives and re-organising how road reserves are managed.

This chapter describes an approach being developed to maintain the biodiversity on road reserves, an approach that is both practical and should give security to biodiversity in the long term. This is about a process to put blood back into these ecosystem arteries and thereby help restore a small part of the naturalness of Australian landscapes and the biodiversity they contain.

Management context

The road reserves are managed already. The managers have objectives, processes, organisations and procedures and there is legislation including environmental law (Bates 1997). The stage is full of players moving in different directions but some movement is towards sustainable development and biodiversity conservation. There is a considerable amount of good will towards environmental objectives and road authorities have been cognisant of the issues for some time (see for example VicRoads 1990 and Ellis 1994). The movement to maintain biodiversity, signified by government strategies, is not yet paramount but it may become so (Council of Australian Governments 1996; Victorian Government 1997). Moving on to implementing management that effectively maintains biodiversity is a logical progression from current environmental concerns. This is not going to be achieved by adding another layer of management or by adding another objective and it is unlikely to be achieved by doing what is being done now, but harder and faster. It is likely to require a small shift in many of the technologies we use, especially in management, control, information and research technologies. It is about changing 'how' road reserves are managed.

There seem to be two items that may be delaying the development of biodiversity maintenance management. One is the still-fuzzy objectives and responsibilities concerning biodiversity and the other is the lack of an effective management system. Once agencies responsible for road reserves accept that maintaining biodiversity is an absolute necessity they will probably turn their attention to cost-effective ways of discharging this task. The task is especially onerous because of the large number of players involved, the dearth of reliable management information, the geographic spread of the reserves and the lack of knowledge and training facilities in biodiversity maintenance practices. Responsible agencies may come to the conclusion that maintaining biodiversity on road corridors is going to be a substantial and ongoing business.

The management approach presented here is still being developed. It starts with the idea of forming locally based co-operatives supported by industry and government. Implementation of this approach will alter what happens on road reserves and will be reflected eventually in what happens in legislation, research, information, communications, education, responsibilities, funding and organisational arrangements. The approach is called the National Protocol System (NPS) and is a management process for identifying and maintaining biodiversity on road reserves and also in adjacent ecosystems (Farmar-Bowers 1998). The development of the NPS has been funded by Austroads, the Association of State Road Authorities.

National Protocol System

Establishing a comprehensive management arrangement at the outset is an investment that will save time, effort, money and biodiversity by helping to ensure individual actions can be co-ordinated as part of a complete system and made effective. The National Protocol System (NPS) will ensure that work will not be lost. For example, a detailed botanical study of a section of road reserve can be designed and executed to the specifications required for the NPS and therefore make an immediate and positive contribution to management practices and so help achieve the NPS's principal objective of maintaining biodiversity. Without the NPS, the study may languish in an agency and may never be interpreted for use by relevant stakeholders except by good fortune. In addition, the very fact that the NPS is functioning will encourage scientific study of road corridors issues since researchers and funding organisations will both know at the outset that the work will contribute to the maintenance of biodiversity.

Who are the stakeholders?

The NPS is a formal co-operative management system involving every stakeholder. Stakeholders are defined as people (agencies or firms) who use the transport reserves or have an interest in the biodiversity they contain or impact. Collectively stakeholders have the responsibility for the past, present and future state of these reserves. The NPS is a system that facilitates and encourages stakeholders to execute their individual and collective responsibilities to maintain biodiversity effectively.

The NPS recognised three groups of stakeholders based on their role in the maintenance of biodiversity. The first group is the site-specific stakeholders; these are people and agencies with an interest in specific road reserves. Site-specific stakeholders include the road authority, the road maintenance contractors, local government authorities, the telephone company, the electricity company, the fire brigade, adjacent landholders, pipeline operators, people collecting seed from native vegetation, owners of grazing stock and so on. Site-specific stakeholders also include people with information on biodiversity or a capacity to get such information. These stakeholders include conservation departments, herbariums, university natural resource departments, several CRCs, students and ecologists. Site-specific stakeholders are the 'doers'. Individually they have the capacity to destroy the existing biodiversity quite easily but if they act collectively they have most of the capacity (but not all) to identify and maintain it.

The second group of stakeholders is industry stakeholders. These are people and agencies that support the site-specific stakeholders with 'industry services'. For example, farmers are site-specific stakeholders and are supported by State governments' primary industry departments and their own organisations (farmers' federation/society). The primary industry department and the farmers' federations are therefore industry stakeholders. Likewise the road authorities have Austroads, advisory committees such as roadside conservation committees and ministerial councils. Local government has Australian Local Government Associations, State local government associations, the Murray-Darling Basin Commission, catchment management authorities and so on. Businesses have employer groups, associations, consultants, standards associations and unions. Ecologists have conservation groups, universities, State conservation departments, museums, professional institutions and so on. Students have schools, colleges, universities and student unions as well as education departments. The industry stakeholders have a support role in biodiversity maintenance; if they wish to they can help site-specific stakeholders by providing technical support, training, encouragement, specific resources and information. They also have the vital capacity to legitimise the role of the site-specific stakeholders in actions to maintain biodiversity. State associations and industry bodies can wield a considerable amount of peer pressure.

The third group is the government stakeholders, specifically the Commonwealth and State governments. They can support the biodiversity maintenance initiatives of the other two groups of stakeholders by providing funds, legislation reform, biodiversity strategies, information, research facilities, co-operative incentives and other supporting actions on the State and national fronts. The NPS could be seen by governments as a vehicle to bring into reality their promises to maintain biodiversity (eg, see Council of Australian Governments 1996).

The NPS is aimed at bringing all stakeholders into co-operative biodiversity maintenance arrangements because each stakeholder has a necessary and complementary role to play. Young has made similar recommendations about the importance of involving all stakeholders in biodiversity conservation programs — see Section 8.5.4 in Young et al. (1996). To be fully effective the NPS requires that stakeholders in each group become effectively involved. For instance, if the local community or farmers decide not to participate no amount of funding will prevent the eventual demise of biodiversity. Individual stakeholders have much more power to remove biodiversity than to maintain it. For example, grassland that developed over thousands of years can be ploughed out in a few hours and the soil/plant relationship cannot be made pristine again.

Some stakeholders can have more than one role and therefore fall into more than one group of stakeholders. For instance, a State road authority can be a site-specific stakeholder for the road system it directly manages and also an industry stakeholder for the road system as a whole.

Co-operation to maintain biodiversity can start in any of these three groups. For example Austroads' current work in biodiversity (funding the development and trial of the NPS and background manuals) is at the 'industry stakeholders' level. Co-operatives of site-specific stakeholders may be the hardest to start because so many people are involved even for a co-operative covering only a few kilometres of road reserve. On the other hand it is possible that co-operatives of site-specific stakeholders once formed may be persistent and vocal in their demands.

Forming site-specific co-operatives may be the key to getting the NPS started.

Site-specific stakeholder co-operatives

The NPS aims to bring all site-specific stakeholders into a single agreement called the Protocol Core which contains one principal and four operational objectives and seeks stakeholder commitment to these objectives through documents called Protocol Chapters (Farmar-Bowers 1995b, 1996). A

Protocol Chapter is like a mini business plan for each site-specific stakeholder. The Protocol Chapter contains both the stakeholder's commitments to the objectives listed in the Protocol Core and a description of how the stakeholder proposes to meet these commitments in practical terms over a specific period of time.

> **Objectives of the National Protocol System**
>
> **Principal Objective of the Protocol Core (draft)**
> *Protection of biological diversity and maintenance of ecological processes and systems on road reserves and in adjacent natural ecosystems, including adjacent waterways.*
> **Operational Objectives of the Protocol Core (draft)**
> 1 *Devise and implement a co-operative management system.*
> 2 *Develop and disseminate knowledge through the co-operative management system*
>   *2.1 Document roadside biodiversity*
>   *2.2 Document threatening processes*
>   *2.3 Improve management information.*
> 3 *Involve all stakeholders in biodiversity management decisions.*
> 4 *Establish independent stakeholder review systems.*

There would be one document called the Protocol Core and each stakeholder would produce their own Protocol Chapter, so there would be hundreds of chapters. For a specific area of road reserve however, only a few stakeholders would be involved and only their chapters would be relevant. The system is manageable but it will rely on the use of computers to access protocol chapter information based on a geographical information system (GIS). The protocol core and chapters are public documents.

Operational Objective No. 4 in the Protocol Core — Establish independent stakeholder review systems — is very important as it provides for a quality control system to be established through agreement by all stakeholders at the time the system is established. It is anticipated that a management committee of stakeholders established by the site-specific co-operative would govern the independent stakeholder review system.

The Protocol Core and Protocol Chapters are the first two elements of the NPS and represent agreement and commitment to act in a specific way. The third element in the NPS is 'Management Arrangements'. Management Arrangements are all the actions and resources needed to allow these commitments to proceed. Management Arrangements include laws, funding, information systems, procedures, research and site management information.

The NPS is an integrated system and so it will not be effective unless it is adopted as a complete system: (1) objectives, (2) commitment to the objectives and (3) the capacity to proceed. Some parts of the Management Arrangements such as the information system and development of reliable management programs may take decades to become fully operational yet they need to be initiated at the outset.

The NPS is a formal system that requires the eventual inclusion of all site-specific stakeholders. An arrangement that allows stakeholders to work together easily is needed. The form of this arrangement has not yet been determined in a practical situation. The establishment of a co-operative as a legal entity has merit but it was not possible to trial such an arrangement in the 1998 program. The nature

Table 14.1   Work under the National Protocol System funded by Austroads as a contribution to the NPS's 'Management Arrangements'.

Items	What is it	Output**
Information system	GIS-based management information, including stakeholders' names and their Protocol Chapters	Working Document drafted
Jurisdiction surety	legal responsibility for biodiversity	Manual: *Environmental Law and Road Reserves*: Bates 1997 (second edition in press)
Decision process	why we undervalue biodiversity	Humphries 1996
Site management*	all the scientific management information we need	McRobert 1997 (road drainage)

* Site management is a huge topic covering every aspect of maintaining biodiversity in perpetuity.
** These reports are available from ARRB TR Ltd, 500 Burwood Hwy, Vermont South, Vic. 3133.

of the co-operative arrangement between site-specific stakeholders may be crucial for the ongoing flexibility of the NPS. It may also alter how some stakeholders view their obligations to the objectives of the NPS and to other stakeholders. A formal co-operative may provide the necessary stability for encouraging serious long-term commitment. It would be useful to employ a management specialist to investigate the options and published the results.

Once site-specific stakeholders start working on maintaining biodiversity, they will be faced with a similar range of issues. The NPS helps site-specific stakeholders address these issues co-operatively. The initial opportunity to co-operate occurs for site-specific stakeholders when they review and approve each other's Protocol Chapters. Industry stakeholders can contribute at any time by providing information or undertaking relevant research. Table 14.1 lists work funded by Austroads as part of their contribution to the NPS.

By developing the NPS, stakeholders will have a system in which they can devise, develop, implement and fund issues relevant to managing the maintenance of biodiversity to a degree that is not possible at the moment. This improved effectiveness will come by firstly pooling existing resources and effort by site-specific stakeholders and secondly by establishing links and joint programs with industry and government stakeholders. The NPS will include an information distribution system to get research and 'how-to management information' to the people working in the reserves.

Most larger road authorities, electricity companies and pipeline operators are already developing and using geographic information systems as part of their asset management control. The information needed for the NPS is substantial and using GIS could be very effective. Operatives such as road authorities do not need all the scientific information on which management instructions may be based. They require only site management information, including information on other stakeholders and their programs. They need to know what to do, whom to consult for advice and how to record their activities. Other stakeholders will require the scientific data and so need access to the other stakeholders' databases. The cost of running these systems is substantial but the availability and power of the information systems in the co-operatives will greatly influence the effectiveness of the work programs. Co-operation and goodwill between stakeholders about sharing information is very important. The additional cost as a consequence of operating co-operatively and having the capacity to make information available to other stakeholders may be relatively small but should be factored into the design of the systems used.

Maintaining biodiversity is a different objective to the conservation of aesthetically pleasing landscapes or the conservation of particular species. It is about conserving in situ all native taxa, most of which we know little or nothing about. Many locations that are important for biodiversity may not be aesthetically attractive and the stakeholders, even if well versed in biology, will need the help and advice of specialists. Without increasing knowledge and information on how to apply that knowledge, planners and site managers will not be able to maintain biodiversity even if they are fully committed to sustainable development ideas. The specialist advice can be created through sophisticated research and investigations. This is the sort of activity that arrangements such as the NPS can facilitate.

Once the NPS is running, the stakeholders themselves would control what research needs to be done collectively on a national or regional basis. The need for firmer links between researchers and those who will ultimately use the results is important (see Specific Recommendation 11.2 in Young et al. 1996). Stakeholders may also have access to various funding arrangements, depending on the kinds of research being proposed. The stakeholders themselves would devise the aim of this research; it might be something like: 'to constantly improve the information available to maintain biodiversity and develop products biodiversity managers require (such as improved information systems)'. Although ideally all stakeholders should be involved in this process, the great disparity between stakeholders may lead to the larger stakeholders not only providing the funds but also selecting the research and development programs.

Overall the NPS would be run, developed and improved by stakeholders themselves. The NPS is neither a 'top down' nor 'bottom up' approach. It concerns stakeholders working together directly to put into effect their joint responsibility for biodiversity maintenance on these strips of public land.

The NPS and adjacent ecosystems

The purpose of the site-specific co-operative is to maintain the biodiversity that occurs on transport reserves. But the stakeholders can also use the road reserves to protect biodiversity in adjacent ecosystems. The issues include (1) drainage water leaving the transport corridor and altering the adjacent ecosystems, especially the aquatic ecosystems and (2) the roads altering the adjacent ecosystem by weed contamination and micro-climatic changes. The drainage issue is as important in urban areas as it is in rural areas because all drainage can impact aquatic biodiversity. This makes the NPS a useful arrangement for dealing with urban road drainage issues.

Austroads acting as an 'industry stakeholder' is currently funding the development of a manual on 'how to drain roads yet maintain biodiversity' (McRobert and Sheridan in press). Previously Austroads also funded a preliminary report on drainage management (McRobert 1997). Both studies are under the auspices of the NPS as a contribution to 'Management Arrangements'.

NPS principles

The NPS is based on a number of principles that are congruent with the objectives of the National Strategy for Ecologically Sustainable Development and the National Strategy for the Conservation for the Conservation of Australia's Biological Diversity.

The twelve principles are summarised below. These are still in draft form.

Principle 1   All stakeholders must have the opportunity for effective involvement.

Principle 2   Site management activity must be a step towards maintaining biodiversity and agreed by the majority of stakeholders as moving towards biodiversity maintenance.

Principle 3   Information must be available to all stakeholders. Information must be relevant for day to day management.

Principle 4   There must be a way of keeping the focus on maintaining biodiversity.

Principle 5   There must be an agreed program aimed at improving the quality of site management practices and the biodiversity maintenance knowledge of people working on linear reserves.

Principle 6   There must be a open, independent review system of (1) stakeholders' intentions (their work programs and plans set out in their Protocol Chapters) and (2) the outcomes of their work in the field.

Principle 7   Back-up institutions must be developed (such as laws, taxonomy, funding, training, scientific and management research).

Principle 8   The system may be started in a region by any stakeholder or group of stakeholders but as a co-operative system it is not to be dominated by interests other than the public interest of biodiversity maintenance.

Principle 9   There must be ways of effecting compliance.

Principle 10   The work and funding for biodiversity maintenance must be apportioned fairly between stakeholders.

Principle 11   There should be effective incentives to achieve and maintain biodiversity.

Principle 12   New ideas and complaints must be dealt with in a fair way.

Getting started

Establishing a co-operative system involving a number of parties is unlikely to happen easily (Farmar-Bowers 1997). A catalyst may be needed to pull the co-operative into existence. The catalyst could be a stakeholder or a group of stakeholders. It could also be an independent broker, perhaps funded externally from government or by an industry association/federation. Whatever the arrangement chosen, the catalyst would broker the network arrangement and move on once the co-operative system is running. The NPS is about stakeholders acting together to achieve the maintenance of biodiversity. While a group of stakeholders may employ a broker to make it more effective, it ceases to be a co-operative when stakeholders no longer participate directly but delegate power to an executive.

Once in existence, the system should constantly provide stakeholders with an effective way of meeting their biodiversity maintenance objectives. If it is not self-maintaining it probably means that some of the stakeholders have decided that they do not need to be involved any longer or they originally agreed to get involved for reasons other than to maintain biodiversity.

There are several steps needed to get the NPS started:

1 Select a person or group to stimulate interest in road reserve biodiversity maintenance with other stakeholders.

2 Select a site such as a region, a local government area, or section of road as the subject matter for the NPS. Meet with most of the site-specific stakeholders and the main industry stakeholders.

Get an idea of the biodiversity maintenance requirements for the site in at least general terms. This may require a flora and fauna study.

3 Reach agreement with the site-specific stakeholders to initiate the NPS. Broker a co-operative arrangement with the stakeholders.

4 Have the site-specific stakeholders indicate their contribution to working co-operatively by preparing Protocol Chapters and develop a site maintenance program based on this information. Approach industry and government stakeholders for resource support.

5 Establish management arrangements such as (1) an information service, (2) site management investigations and (3) a monitoring and auditing system including a formal independent review system.

6 Develop a fund raising and expenditure program that is equitable.

7 Develop a program to regularly check the objectives of stakeholders to make sure they are compatible with maintaining biodiversity.

Once the NPS is running, the geographic areas it covers can be expanded and the involvement of stakeholders consolidated.

Starting a co-operative will require involvement of several stakeholders but not every one. Table 14.2 indicates their possible roles.

Performance of the NPS

The bottom line is whether or not the co-operative venture delivers the principal objective of the Protocol Core (biodiversity maintenance). Effectiveness for the NPS is the identification of biodiversity and its perpetual maintenance. Interim measures of success for the co-operative might include reaching high levels of funding, research, biodiversity surveys, technology transfer and so on, but the ultimate measure is being effective in biodiversity maintenance. Getting site-specific co-operatives and the full NPS up and running may be significant, but only as management tools to get comprehensive biodiversity surveys and research programs underway and the results implemented.

Table 14.2   Roles in co-operative arrangement for biodiversity conservation.

Role of				
Initiation group or person	Lead stakeholder	Broker	Site-specific stakeholders	Industries & Governments
To lobby the stakeholders into action	Promote a co-operative with other stakeholders	Broker a co-operative agreement with all site-specific stakeholders	(1) Commit to the Protocol's Core Objectives in a Protocol Chapter and form a co-operative (2) to devise a biodiversity plan for a site (3) to execute the plan co-operatively	Support site-specific stakeholders' co-operatives with funds, management information, GIS, legislation reform and strategies

Stakeholders need assurance that their co-operative actions are actually maintaining biodiversity. Appreciating what actions are actually maintaining biodiversity is not easy. Gross actions like clearing land or introducing exotic plants obviously damage biodiversity but, even with the best of intentions, well meaning actions can lead to biodiversity loss. There is very little reliable research-based information on how to manage linear reserves for the best long-term results. It may take years of research, trials and observation before the body of knowledge on 'what to do to maintain biodiversity' on road reserves becomes comprehensive and reliable.

In the meantime, stakeholders could start accurately recording their current management action on a GIS basis. Such information will be invaluable later on for helping to determine the best management to maintain biodiversity.

Developing realistic performance indicators is difficult. Three simple qualitative ones that might be useful for a time are suggested below. Measures of progress can be invaluable guides for management but the benefit needs to be weighed against the effort involved in establishing and using performance indicators and also in the possibility of focusing resources on achieving an inappropriate indicator.

Perhaps the first performance indicator at this time could be 'arrangements that focus a site-specific co-operative exclusively on biodiversity conservation'. At the same time the co-operative could be making sure that there are no arrangements that detract from that focus. Perhaps the most likely detractions could be 'environmental issues' such as landscaping, tree planting, forms of erosion and pest control, and aesthetics.

The second performance indicator could be 'administrative arrangements that keep interest levels high and the co-operative alive and moving'. Arrangements could ensure the people involved in biodiversity maintenance get approval and recognition from the highest levels in our society. It may be possible to build into the NPS a national review program though techniques like special biodiversity publications, network groups, national conferences, university research scholarships, school programs and awards.

Effectiveness in maintaining biodiversity should be the principal determinant of stakeholder satisfaction in the co-operative. The NPS is founded on sustainable development ideas. The 'welfare and well-being objective' of sustainable development relies on participation for achievement. Participation and having control over one's own affairs is an important aspect of sustainable development (see Farmar-Bowers 1994). The NPS offers a system that allows participation. Perhaps the 'level of participation' could be used as a third performance indicator.

The three qualitative performance indicators are thus 'focus', 'activity' and 'participation'. The NPS is probably being effective if a lot of people participate and are co-operatively very active with a focus on biodiversity maintenance.

Trial of the NPS

The NPS is being trialed in two regions in Tasmania during 1998. This trial is being funded jointly by Austroads and the Department of Transport, Tasmania. Reports on progress will be available through the Tasmanian Department of Transport and Austroads sources (Farmar-Bowers in press).

Two two-day workshops were held with stakeholders (mainly site-specific stakeholders) and a program of action devised. The workshop for the second region was preceded by a fauna and flora survey that identified specific biodiversity maintenance issues in a 60-kilometre section of highway.

The next phase of the trial will involve bringing together stakeholders who did not attend the workshops and working with all involved to establish commitment to biodiversity maintenance via Protocol Chapters, then having the stakeholders get a site management program developed, agreed and operational. As time progresses the opportunities to expand the trial in geographic areas and with other groups of stakeholders elsewhere will be pursued. The work will be integrated with other conservation-oriented work in the catchment. The author is acting as the broker in this trial.

Interaction with other conservation programs

The common element between the NPS and other conservation work is the 'site management practices'. For example, a program to protect aquatic habitat from sedimentation is applicable whether the source of sediment is coming from farmland, national park or a road reserve. The link with conservation groups is also through the mutual need for reliable information and programs on how to maintain biodiversity and ecological processes.

The focus of some conservation programs may not be biodiversity maintenance; it may be aesthetics, weed control, salinity control, agricultural, integrated catchment management, environmental impact statement, archaeological or cultural. The connection with biodiversity maintenance would invariably be in site management works — what these programs deliver 'on the ground'. Conservation programs may assist in maintaining biodiversity on transport reserves even if their focus is somewhat different.

Where the adjacent land is owned and managed by a single entity such as a national park or a farm, a co-operative management system like the NPS is not needed as conservation programs can easily be negotiated with the manager of the land. Things become more complex as the number of agencies and individuals (stakeholders) with access to the land increases. The risk is that one of the stakeholders will nullify the conservation program because they have not been informed or it is not in their interests to comply. However there may be advantages in having to deal with so many stakeholders because the range of agencies creates a capability of undertaking co-operative study programs in 'site management practices'. In this way the effective operation of the NPS will provide a substantial benefit to neighbours of the road reserves by providing well researched 'how to maintain biodiversity' information. Of course once the NPS is operating, a site-specific co-operative should be able to function as a single entity, like a normal landowner, to take advantage of the range of existing conservation programs.

Some conservation studies are undertaken to satisfy environmental impact assessment needs. Some of the technical work done may assist stakeholders maintain biodiversity. This might be especially so if the study is being done as a contribution to a 'strategic environmental impacts statement' as the timeframe and context may be broader. If a NPS co-operative has a well advanced program in the region it could provide information to help planners meet the requirement of the environmental impact legislation.

Road drainage can impact adjacent ecosystems, especially aquatic ecosystems. Erosion in the drains can lead to sedimentation problems in receiving streams.

Information from programs such as Landcare would be synergistic mainly because the people involved with Landcare are often farmers and likely to become site-specific stakeholders under the NPS. Programs commissioned by catchment management authorities such as roadside management strategies help identify issues and raise awareness (McGuinness 1998). By clarifying the options, such strategies would facilitate the establishment of NPS site-specific stakeholder co-operatives. Essential ingredients of the NPS include (1) quality of information assessed through the peer review system and (2) sharing of information, both of which are compatible with most environmental programs. The NPS is consistent with quality systems such as ISO 14000 series and could be used as an environmental management system.

The NPS is generally consistent with biodiversity strategies. For instance the NPS provides a process or mechanism for achieving the Actions listed in the Draft NSW Biodiversity Strategy (1997) (in particular the Priority Actions Nos. 9, 25, 54, 87, 98, 144, 172 and 195) for road reserves. For road reserves and adjacent ecosystems, the NPS is a model biodiversity planning framework and it could be developed, with a little modification, into a model bioregional planning framework as required in Action 25, which is one of the 14 priority actions of the Draft NSW Biodiversity Strategy.

The cost

Inevitably there is a cost involved in biodiversity. The extremes are 'society pays now by implementing an effective biodiversity maintenance program' or 'future generations pay through living in a country with less biodiversity'. Biodiversity maintenance is an inter-generational equity issue. It seems likely that the end result will be somewhere in the middle with the current generation choosing to pay for some biodiversity maintenance programs and moving the rest of the cost onto future Australians.

Mallee vegetation looks dense to travellers but is actually only a few metres wide.

Trees on a road reserve in western Victoria improve the aesthetics of the road. The understorey is mainly composed of introduced grasses.

There is no simple economic argument possible here because we simply do not know what biodiversity is at stake, let alone its discounted current value. In general terms the rationale may be that biodiversity should be maintained for equity reasons and that we should be looking for the most economic way of achieving this. Working together and adapting the existing institutions to the task may the cheapest effective way forward (which is the NPS).

Funding is a very sensitive issue but the gathering of funding is related to responsibilities. A view of the responsibility

A temporary measure to control a permanent erosion-sedimentation problem in New South Wales.

involved could be that in general, the maintenance of biodiversity is a national issue. This would indicate that the base funding for maintaining biodiversity is a collective responsibility of all Australians, irrespective of where that biodiversity is situated. An additional view is that transport corridors provide special value to those who use them. This value is mainly associated with the space in reserves needed for facilities such as road pavements, drainage systems, pipelines, power cables and so on. Once this space is used for facilities, it is no longer available to support native ecosystems. The road reserve users therefore have a responsibility for biodiversity in addition to the general one that all Australians have. The responsibility for providing funds for biodiversity maintenance on transport corridors could therefore be split between all Australians (via general tax revenue) and those Australians who use the transport corridor (funds provided by transport agencies and firms with infrastructure on the transport corridor).

It would be unfair for local people to have to pay for biodiversity maintenance just because their region happened to retain significant biodiversity. In addition this may be a disincentive for biodiversity maintenance as they may be tempted to save money by destroying biodiversity. There may be some value in resurrecting the concept of 'biodiversity bonds' initially discussed by Farmar–Bowers (1995a). This was an idea that would allocate value to areas of land supporting native vegetation and animals and allow tradeoffs between losing some native areas of vegetation in return for resources

Vegetation on the road reserve is vulnerable to many small changes but rarely survives the decision to widen the pavement.

options such as substitute land, research or other requirements for biodiversity maintenance.

The total amount of funds gathered will partly determine the split between paying for biodiversity maintenance and letting future generations suffer the loss.

How the money is spent is also an equity issue. Equitable expenditure requires that the money be spent effectively on biodiversity maintenance. The objective is to give the funds to those people who can achieve the objective. For example, if the program involved conserving wetlands and there is not enough information to identify species, then funds in this case should be diverted to taxonomic groups such as herbariums to ensure the information is developed in time. The objective is overall effectiveness, and 'weak links' in the effectiveness chain need to be identified and addressed even if that involves funding agencies and groups not normally funded through the transport industry. For example, the road agencies, electricity and telephone companies may have to fund State herbariums and university ecology research to get the ongoing knowledge they need. Local landholders frequently benefit from using the road reserve; their contribution to biodiversity maintenance may be 'work in kind' — this is raising and expenditure of funds simultaneously.

In the longer term, the resources needed to maintain biodiversity may be considerable and require the involvement of State and Federal governments. Initially the NPS may be able to start by using funds provided by industry and site-specific stakeholders but eventually the general population may have to contribute via State and Commonwealth funding.

Conclusion

A road traversing forest in New South Wales. Roads can alter the adjacent forest in a number of ways, notably by acting as a conduit for weeds and pest animals.

Road reserves contain important remnants of native ecosystems and can also be used to protect adjacent ecosystems. These remnants are vulnerable to loss. The management of road reserves that achieves biodiversity maintenance will make a significant contribution to the overall conservation of Australia's biodiversity.

Current management practices could be improved gradually by harnessing the goodwill of the many people involved. To ensure that their efforts are made effective, the author is proposing the adoption of the National Protocol System which is a formal co-operative management system for biodiversity maintenance that involves every stakeholder. The focus of the system will be kept on biodiversity maintenance by including all stakeholders and having the stakeholders themselves run an independent review system.

The results of the trial of the National Protocol System in Tasmania being undertaken in 1998, if positive, may lead to the further development of the NPS and its wider use in other areas in Australia.

References

Bates, G. (1997), *Environmental law and road reserves*. Special Report No 55. (ARRB Transport Research Ltd: Melbourne.)

Breckwoldt, R. *et al.* (1990), *Living corridors*. (Greening Australia: Canberra.)

Commonwealth of Australia (1996). *Reimbursing the future — an evaluation of motivational, voluntary, price-based property-right and regulatory incentives for the conservation of biodiversity.* Biodiversity Series, Paper No. 9. (Biodiversity Unit, Dept of Environment Sport and Territories: Canberra.)

Council of Australian Governments (1996). *National strategy for the conservation of Australia's biological diversity.* (Commonwealth Department of Environment, Sport and Territories: Canberra.)

Ellis, M. (1994). *Roadside Management Manual.* (Shire of South Gippsland and Woorayl: Victoria.)

Farmar-Bowers, Q. (1994). *A self-trainer in ecologically sustainable development analysis.* Research report. (Australian Road Research Board Ltd: Melbourne.)

Farmar-Bowers, Q. (1995a) *Is there a road industry responsibility for biological diversity?* WD RS 95/002. (Australian Road Research Board Ltd: Melbourne.)

Farmar-Bowers, Q. (1995b). *First draft, national protocol core, national protocol system, biodiversity conservation on roadsides and in adjacent waterways.* Sustainable Development Series, TO WD 95/008. (ARRB Transport Research Ltd: Melbourne.)

Farmar-Bowers, Q. (1996). *Protocol chapters, national protocol system, biodiversity conservation on roadsides and in adjacent waterways.* Sustainable Development Series, TO WD 96/004. (ARRB Transport Research Ltd: Melbourne.)

Farmar-Bowers, Q., (1997). Implementing The National Protocol System Down Under: Cooperative Management Device for Biodiversity Conservation on road corridors in Australia. In *6th International Symposium, Environmental Concerns in Right-of-Way Management, 24–26 February 1997, New Orleans, USA.* (Eds J.R. Williams, J.W. Goodrich-Mahoney, J.R. Wisniewski and J. Wisniewski.) pp. 375–382. (Elsevier Science: Oxford.)

Farmar-Bowers, Q. (1998). *The national protocol system in 1997: a cooperative management system to maintain biodiversity.* Research Report, ARR 317. (ARRB Transport Research Ltd: Melbourne.)

Farmar-Bowers, Q., (in press), *Managing the Maintenance of Biodiversity on Road Reserves: Report on the Trial of the National Protocol System, Tasmania 1998,* (incorporating the final report, the two progress reports and 11 Newsletters) Prepared for Austroads, Department of Transport, Hobart, Tasmania.

Humphries, E. S. (1996). *Enhancing the human element in environmental decision making.* Sustainable Development Series No. 7, TO WD 96/037. (ARRB Transport Research Ltd: Melbourne.)

Huston, M.A. (1994). *Biological diversity, the coexistence of species on changing landscapes.* (Cambridge University Press: Cambridge, UK.)

McGuinness, S. (1998). *Roadside management strategy for the mallee catchment.* (Commissioned by the Mallee Catchment Authority, McGuinness and Associates, Bendigo, Victoria.)

McRobert, J. (1997). *Biological diversity in transport corridors: road drainage management.* Research report ARR 302. (ARRB Transport Research Ltd: Melbourne.)

McRobert, J., and Sheridan, G., (in press 1999). *Road Runoff and Drainage: Environmental Impact and Management Options, First Draft Manual* (ARRB Transport Research Ltd.: Melbourne.)

NSW Government (1997). *Draft NSW biodiversity strategy.* (National Parks and Wildlife Service: Hurstville, NSW.)

Reaka-Kudla, M.L., Wilson D.E., and Wilson E.O. (Eds) (1997). *Biodiversity II: understanding and protecting our biological resources.* (Joseph Henry Press: Washington DC.)

Saunders, D.A., and Hobbs, R. J. (Eds). (1991). *Conservation 2: the role of corridors.* (Surrey Beatty & Sons: Chipping Norton, NSW.)

Victorian Government (1997). *Victoria's Biodiversity, Directions in Management.* (Department of Natural Resources and Environment: East Melbourne, Victoria.)

VicRoads (1990). *Roadside management guide.* General report /90-6. (VicRoads: Kew, Victoria.)

Young, M.D., Cunningham, N. Elix, J., Lambert, J., Howard, B., Grabosky, P. and McCrone, E., (1996). *Reimbursing the Future,* Part 1, Biodiversity Series, Paper No 9, Biodiversity Unit, Department of the Environment Sport and Territories, Canberra.